D0984224

FISH CHROMOSOME METHODOLOGY

FISH CHROMOSOME •METHODOLOGY•

By

THOMAS E. DENTON
Associate Professor of Biology
Samford University
Birmingham, Alabama

CHARLES C THOMAS • PUBLISHER
Springfield • Illinois • U.S.A.

Published and Distributed Throughout the World by

CHARLES C THOMAS • PUBLISHER
BANNERSTONE HOUSE
301-327 East Lawrence Avenue, Springfield, Illinois, U.S.A.

ISBN 0-398-02831-1

Library of Congress Catalog Card Number: 73-2688

With THOMAS BOOKS *careful attention is given to all details of manufacturing and design. It is the Publisher's desire to present books that are satisfactory as to their physical qualities and artistic possibilities and appropriate for their particular use.* THOMAS BOOKS *will be true to those laws of quality that assure a good name and good will.*

Library of Congress Cataloging in Publication Data

Denton, Thomas E
Fish chromosome methodology.

Includes bibliographical references.
1. Fish genetics. 2. Karyotypes. I. Title.
QL639.D46 597'.01'51 73-2688
ISBN 0-398-02831-1

Printed in the United States of America

N-1

PREFACE

SINCE 1960 A VAST amount of interest has been shown in the area of fish cytogenetics. The increase in knowledge in this area is attributed to improved methodology for obtaining and studying chromosomes from fishes. Many of the methods are offshoots from mammalian cytogenetics although some have been developed especially for studying fishes and amphibians.

This book is a fundamental guide to studying fish chromosomes. The major portion of material is taken from contributors in the field of fish cytogenetics within the past fifteen years. Although the book is written primarily for the beginning researcher, it is intended that the veteran investigator also benefit from its contents. The six chapters are divided into three general areas: Classification and maintenance of fishes; obtaining and presenting chromosome data; and a listing of fish chromosomes with an analysis of the fish karyotype as it is currently known. With this approach it is felt that the researcher may be better assisted in presenting and evaluating his findings relative to the works of other investigators in the area of fish chromosome study.

I would like to thank all of the scientists who commented and offered suggestions concerning the presentation of their methods and to Dr. Earnest DuPraw for supplying me with an electron photomicrograph of a representative human chromosome. I owe a special debt of gratitude to my colleagues, Dr. H.A. McCullough and Dr. William Mike Howell, who carefully read the manuscript and suggested valuable improvements.

T.E.D.

CONTENTS

FISH CHROMOSOME METHODOLOGY

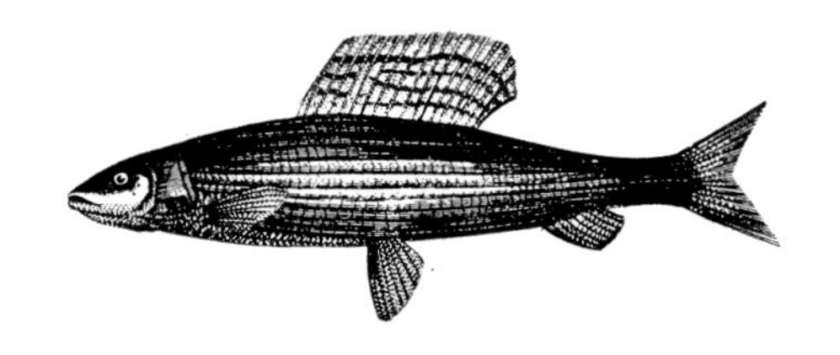

CHAPTER 1

INTRODUCTION TO WORKING WITH FISHES

ANY TREATMENT OF the study of fishes is a vast undertaking. They are a numerous and diverse group of organisms with varied morphology and physiology, complex patterns of distribution, and poorly understood heredity. The emphasis of this text is cytogenetics, which is but one aspect of the world of fishes. Because of this specificity, the writer feels obligated to present some background information about fishes in general with the premise that anyone working in the area of fish cytogenetics may better understand the chapters that follow as well as the context of his work.

A BRIEF SURVEY

From a consensus of estimates, there are approximately 20,000 species of fishes living in the world today. This is almost half of all existing vertebrates with amphibians numbering about 2,500, reptiles 6,000, birds 8,600, and mammals 4,500 (Lagler *et al.*, 1962). There are many more fishes that are undescribed and have caused some estimates to range as high as 40,000. When one considers the depths of the oceans and remote areas of the world that have never been investigated, it is wondered if such high estimates are not justified. Estimations of fish numbers are further complicated by the fact that many populations are becoming extinct. Man, by way of his technological achievements, has unknowingly placed many species of fishes on a rare and endangered list. Anyplace in the world where conditions radically change, can result in loss of fish populations. Conversely, species of fishes are evolving. This is accomplished over a relatively short period of time through hybridization of closely related species that have been brought together by circumstance. The production of new species through adaptation to the environment is a much longer process but one that must be acknowledged as a current happening. When these aspects are considered, it becomes evident that the precise number of fishes will never be known as long as large numbers remain. On the other hand, as man condenses and better defines his global environment, a more accurate figure should eventually emerge.

Fishes exist with a variety of shapes and forms. The one factor altering the forms of fishes more than anything else has been the differential pressures exerted by the water in which they live. These pressures have produced their effects principally during stages of growth and development and have resulted in a diversity of symmetries. Some fishes for example are bottom dwellers and have taken to lying on their ventral sides. These forms have become flattened from above such as the skates and rays which move by undulations of enlarged side fins. Others lie on their sides and have become flattened sideways; e.g., the flounder which lies on its left side with its eyes on the upper right side. Specializations of shape are also determined by feeding behavior. More time and energy is spent on feeding and pursuing food than on anything else. As a result, numerous types of feeding mechanisms with varied structures have developed. Most fishes are generally streamlined or torpedo shaped. This shape is better suited to maneuvering and speeding through the water while being stabilized and steered by fins of various size, shape, and number. Despite these many variations, fishes, as with most vertebrates, are bilaterally symmetrical; the left and right halves are mirror images of each other.

As a group, fishes exhibit great range in size. Some gobies (e. g., *Pandaka pygmaea* of the Philippine Islands) represent the smallest known fish measuring a little over one-third of an inch in length. The largest is the whale shark, *Rhincodon typicus,* which attains a length of over fifty feet and weighs twenty tons or more. The largest freshwater fishes are catfishes and sturgeons. Some Eurasian catfishes have been recorded up to ten feet in length. In the Volga River some sturgeons attain a length of fourteen feet and weigh over two thousand pounds. On the Pacific coast of the United States some sturgeons have been caught that measured over twelve feet. A good example of the diversity of size in fishes of the same kind is exhibited in rays. The smallest ray, located in the South Seas, is less than six inches as an adult whereas sting rays may measure six feet across; and the great black and white devilfish may be twenty-two feet from wing tip to wing tip.

How long does a fish live? This question is difficult to answer. It is possible to estimate the age of a fish by counting scale annuli. In order to do this with some measure of accuracy a proportionality of scale growth to body growth must be determined. Even then, the average longevity for a particular species is hard to determine due to the hazardous environment in which it lives. The life span of fishes living in the wild has been estimated by some to be about twenty to thirty years (Lagler et. al., 1962) . Some small gobies live only a year and fishes that grow longer than a foot are thought to live at least four or five years. Some unauthenticated reports relate that some pond carps have reached ages of two hundred to four hundred years. More confirmed studies, however, suggest that the carp rarely exceeds fifty years. It appears that death comes most soon to the fastest growing individuals and males usually die sooner than females. In the protective surroundings of captivity the life span may be prolonged. The northern pike will live about twenty-five years in nature but up to seventy-five years in captivity.

Approximately 75 percent of the world is covered with water. In this vast aqueous system fishes are living in a variety of habitats. There are an estimated 6,000 species of fishes living in freshwater and some 14,000 constituting saltwater or marine species. Transitions between freshwater and saltwater do not normally occur because of physiological differences between the two fish forms. Some, however, can exist in both (e. g., sticklebacks) or can migrate from saltwater to freshwater (e. g., spawning salmon) or vice versa (e. g., the common European eel) ; and some have become adapted to the inbetween brackish water estuaries along the coasts (e. g., the sailfin molly) . Fishes that are strictly freshwater are confined by land masses. They can naturally pass from one isolated drainage system to another only as a result of a change in the land itself. Marine fishes, on the other hand, are confined by a number of factors the principal one being temperature; i. e., the various temperatures of the global waters act as a barrier and tend to isolate many sea-dwelling fish populations.

It is not clear how the first fish evolved. In fact, it can only be surmised what the first form looked like and how it lived. Somewhere, perhaps five hundred million years ago during the Cambrain period, the first fish-like creature appeared. Figure 1-1 is a suggestive scheme of how fishes evolved over the geological periods of time. The earliest fossil record of a fish form is dated between the Cambrain and Ordovician period about four hundred million years ago. These jawless ostracoderms probably inhabited freshwaters where they lived on the bottom, filtering food through gills that served as both a feeding and breathing apparatus. They had a bony shell, armored back, one pair of rudimentary fins, no paired appendages and in place of a vertebral column, a cartilaginous notochord. The lampreys and hagfishes of today are the only living descendants of these forms. The ostracoderms survived for some 160 million years and died out during the Devonian period. Sometime before this happened, in the upper Silurian, the first fishes with jaws and paired fins appeared. The development of these placoderms of unknown ancestry was a milestone in the evolution of all vertebrates because, with jaws and fins, they could pursue and consume new sources of food. The placoderms became extinct during the Carboniferous leaving no living descendants.

All fishes of today belong to three classes. The lampreys and hagfishes of the class Agnatha have already been mentioned. The other two classes, the Chondricthyes and Osteichthyes, constitute the most successful of today's fish groups. The Chondrichthyes are mostly predatory ocean dwellers and include the sharks, skates rays and chimaeras or ratfish. They are characterized by having a cartilaginous skeleton, small eyes, placoid scales, well developed nostrils, and paired fins. They first appeared in the Silurian and became widespread in the Carboniferous and Permian. After

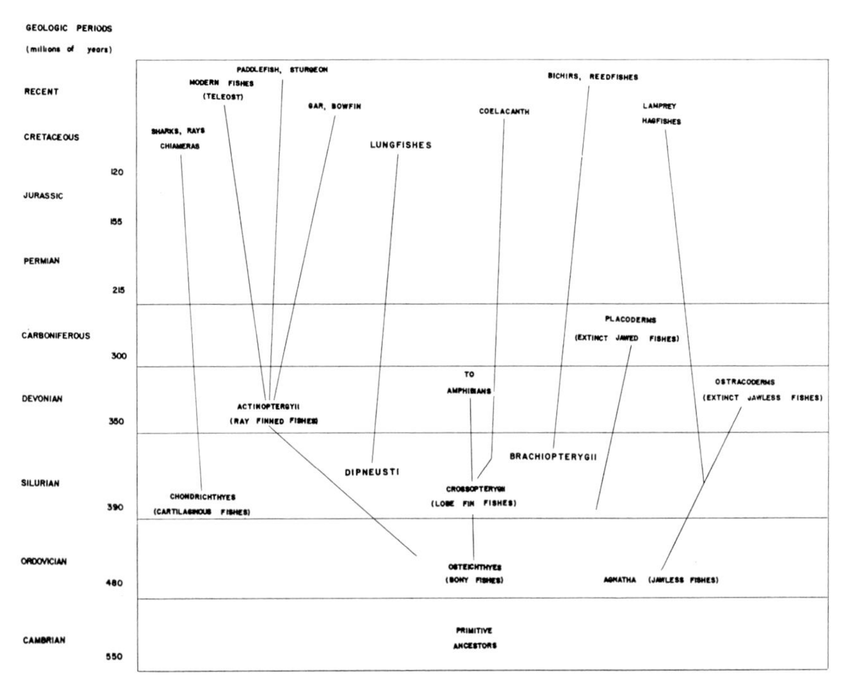

Figure 1-1. A scheme for the evolution of the major fish groups.

some eighty-five million years there was a decline in Chondrichthyan types so that today there is left but a few families of sharks, rays and chimaeras. These living members, however, are very successful and widespread despite the relatively small number of existing species.

The class Osteichthyes, or bony fishes, contains the most numerous of today's fishes. The first member is thought to have appeared in the freshwaters of the Ordovician although fossils do not appear until the Silurian. Since these earliest forms, the bony fishes have radiated into both freshwater and marine environments with the major expansion occurring during the Cretaceous. Unlike the Chondrichthyes, the bony fishes have four pairs of gills in a common cavity, an air bladder, cycloid or ctenoid scales and a more well developed brain. The Osteichthyes are divided into four subclasses. These are the lungfishes, Dipneusti; the lobefins,

Crossopterygii; the bichirs and reedfishes, Brachyopterygii; and the ray-finned fishes, Actinopterygii. The lungfishes are composed of five freshwater species which are confined to Australia, Africa, and South America. They probably arose from an ancient form that also gave rise to the Crossopterygii. All the lobefin fishes were thought to be extinct until a modern representative, *Latimeria chalumnae,* was collected off the coast of Africa in 1938. This is still the only known living representative and constitutes a monotypic genus. The bichers and reedfishes are found in West and Central Tropical Africa. There are twelve species of freshwater bichirs *(Polypterus)* and only one species of reedfish *(Calamoichthyes calabaricus).* It is still disputed where and how these forms originated. The fourth subclass, the Actinoptergyii, is divided into the Chondrostei (sturgeons and paddle-fishes), Holostei (gars and bowfin), and the Teleostei (modern bony fishes). The teleost represent the end group of the ray-finned fishes and are the most numerous and successful of all living fishes.

CLASSIFICATION

One of the basic problems in studying fishes is classifying them into groups. Fish classification, as with most classifications, categorize living specimens according to similarities and differences of systems. This includes physiological, morphological, embryological, and genetical characteristics. In addition, fossil studies determine relationships in non-living forms. As a consequence of this approach an artificial system that is, a practical system, exists because of its simplicity and the unavailability of information on many existant specimens. An ideal system would be one that shows phylogenetic relationships, or the order of evolution of individuals within a group. Such a complete system is not in use at the present time although authorities are constantly striving toward this end. Because of this dichotomy of what is desired and what is practical, ichthyologists disagree in their attempts to classify fishes and show their proper relationships. To the investigator who is not a taxonomist, this endeavor is especially confusing. There are a number of good classification schemes available, but only one will be presented to show what is considered to be an acceptable phylogenetic arrangement between the major fish groups. More detailed and employable classification keys may be found elsewhere (Blair *et al.*, 1968; Breeder, 1948; Eddy, 1969, and Sterba, 1962). The proposed listing that follows is one that is basic and widely used. The classes Agnatha and Chrondrichthyes are those of Berg (1947) as given by Lagler, *et al.* (1962). The subclasses of Osteichthyes are arranged according to Grassé (1958), and the teleost orders are those suggested by Greenwood, *et al.* (1966).

Phylum Chordata
- Subphylum Vertebrata
  - Superclass Pisces: Fishes with gills and fins; body covered with scales; all aquatic; approximately 20,000 species.
    - Class Agnatha: Jawless fishes, without scales or paired appendages.
      - Subclass Cyclostomata: A primitive, jawless, eel-like vertebrate. Among earliest fossil vertebrates but unknown themselves as fossils.
        - Order Myxiniformes: Hagfishes, all marine, found principally in the cool parts of the Northern and Southern Hemispheres. Fifteen species.
        - Order Petromyzontiformes: Lampreys, found in cool zones of both Northern and Southern Hemispheres; about eight genera constituting twenty-four species. There are fresh-water forms but none entirely marine since fresh-water is needed for breeding.
    - Class Chondrichthyes—Cartilaginous fishes with five to seven pairs of separate gills; scales tiny and non-overlapping.
      - Subclass Elasmobranchii: About 550 species.
        - Order Squaliformes: Shark-like marine fishes numerous since the upper Devonian.
        - Order Rajiformes: Marine skates and rays; derived from the sharks during the Mesozoic.

Subclass Holocephali

Order Chimaeriformes: Rare, oceanic. Chimaeras presumably derived from the early sharks.

Class Osteichthyes—Bony fishes with well developed cranium and ribs; gills under gill covers; scales generally overlapping.

Subclass Dipneusti: Lungfishes, all fresh-water, first evident in the Middle Devonian. Now confined to Australia, Africa and South America. Five species.

Subclass Crossopterygii: The lobefins; mostly fossil except for the coelacanth *Latimeria chalumnae* known only in the sea off the southeast coast of Africa.

Subclass Brachiopterygii: The twelve species of fresh-water bichirs and one species of reedfish are confined to West and Central Tropical Africa.

Subclass Actinopterygii: The ray finned fishes; the dominant form of fishes since the Carboniferous. Approximately 19,000 species.

Superorder Chondrostei

Order Acipenseriformes: Sturgeons and paddlefishes; freshwater and marine.

Superorder Holostei

Order Amiiformes: The fresh-water bowfins now confined to East Central North America; probably dominant during the Jurassic.

Order Lepisosteiformes: Fresh-water gar; four or five species of heavily armored fishes; dominant in Eastern North America.

Superorder Teleostei: The end group of the ray finned fishes. Most dominant of fishes in the world today. Presumably originated in the sea during the Jurassic.

Order Elopiformes (Targon and deep sea bone fishes, all marine)

Order Anguilliformes (eels, fresh-water and marine)
Order Notacanthiformes (spiny eels, all marine)
Order Clupeiformes (herring and shad, fresh-water and marine)
Order Osteoglossiformes (bony tongues and mudskippers, fresh-water and marine)
Order Mormyriformes (gymnarchids and mormyrids, fresh-water and marine)
Order Salmoniformes (salmon, trout, fresh-water and marine)
Order Cetomimiformes (deep sea ateleopids and giganturids all marine)
Order Ctenothrissiformes (monotypic marine family Macristiidae)
Order Gonorychiformes (milkfishes and African pygmy fishes fresh-water and marine)
Order Cypriniformes (minnows suckers etc. fresh-water and marine)
Order Siluriformes (catfishes fresh-water and marine)
Order Percopsiformes (perches, all fresh-water)
Order Batrachoidiformes (toadfishes, all marine)
Order Gobiesociformes (clingfishes, all marine)
Order Lophiiformes (goosefishes, batfishes, all marine)
Order Gadiformes (codfishes and rat tails, all marine)
Order Atherinformes (livebearers, mullets, etc., fresh-water and marine)
Order Beryciformes (beard fishes, squirrel fishes, all marine)
Order Zeiformes (dories and bear fishes, all marine)
Order Lampridiformes (ribbon fishes, all marine)
Order Gasterosteiformes (sticklebacks, sea horses, etc., fresh-water and marine)
Order Channiformes (snakeheads, all marine)
Order Synbranchiformes (swamp eels, fresh-water)
Order Scorpaeniformes (rockfishes, sculpins, fresh-water and marine)
Order Dactylopteriformes (batfish, all marine)
Order Pegasiformes (sea moths, all marine)

Order Perciformes (snooks, snappers, fresh-water and marine)
Order Pleuronectiformes (flounders, soles, fresh-water and marine)
Order Tetraodontiformes, spikefishes, puffers, all marine)

THE PROCUREMENT AND MAINTENANCE OF FISHES

It has previously been emphasized that fishes are global in distribution. For this reason problems exist for the investigator who wants to work with species not normally found in local habitats. Provided he is fortunate enough to obtain specimens locally, or is forced to get them from distant places, there is the additional problem of keeping them healthy. This is a prime criterion for the fish cytogeneticist. It is imperative that fishes be healthy and metabolically active if good chromosome preparations are to be obtained from their tissues. Generally, it is best to enumerate chromosomes from a specimen as soon as possible after it has been obtained. This better insures that the tissues will be fresh and healthy, and also reduces the possibility that the organism will die from the many causes that become prevalent when a fish is kept in captivity. Invariably, it is not always feasible to process tissues immediately. It then becomes necessary to make preparations to house the organism for an indefinite period of time. Before these conditions of maintenance are discussed, the fishes must first be obtained.

Collecting

Freshwater fishes are usually collected in the field with nets of various types and sizes or with hook and line. A one man seine (3 ft. × 4 ft.) or a two man seine (10 ft. × 4 ft.), made of $\frac{1}{4}$ inch mesh with stout wooden poles tied on each end, is generally used to collect fishes of all sizes where it is possible to wade and maneuver the net along the bottom of the stream. In large streams or lakes where conventional seining is not feasible, a bag seine, gill net, or hook and line may be desirable for getting large specimens, or minnow traps may be used for collecting small specimens.

Local fishermen and biology departments are usually good consultants if one is interested in specific collecting procedures. Regardless of the technique that is used for collecting specimens, it should be emphasized that the collector have the necessary licenses or permits to comply with state and local regulations in regard to taking fishes. Once they are captured, a styrofoam container filled with stream water is adequate to transport them back to the laboratory. If the distance is greater than thirty minutes, it would be well to use a portable aerator pump to prevent the fishes from suffocating.

The methods for collecting freshwater fishes may also be used for collecting marine forms. There are additional methods, however, that involve more expense and time. Once they are collected, there is the added difficulty of transporting them so that they arrive at their destination in a healthy state. Because of these difficulties, and for the expense involved, it is usually better to contract with one of the many coastal marine institutes for both the collection of saltwater fishes and for facilities where chromosome preparations can be made while the specimens are fresh.

Purchasing

Both freshwater and marine fishes may be purchased from a variety of places (Appendix). Oftentimes this is the only recourse in obtaining fishes that are not found in local habitats. Usually, the best place to find the most varied fish forms is in a well-stocked tropical fish center. Here, fishes may be purchased and taken back to the laboratory for processing. Usually, fishes bought in this way are sufficiently healthy for making good chromosome preparations. Fish dealers will place special orders for specimens that are not normally stocked. Unfortunately, they are usually ordered according to common names and this introduces some risk in actually getting the desired species. Also, it is generally impossible to determine precisely where the organism was collected. State and federal fish hatcheries and biological stations are also sources for procuring specimens. These agencies are generally cooperative with the conscientious researcher and sometimes will even provide facilities for him to work.

Transporting

Requests for many fishes may be placed directly with out of town suppliers or from a colleague who will collect and send them from another locality. Often, this involves subjecting the fishes to hazardous conditions for an undesignated period of time during transit. Fishes that are sent from one region to another should not be in transit longer than twenty-four hours. This necessitates that they be transported by air freight. Specimens sent this way are normally placed into plastic bags containing water that is close to their native temperature and salinity.

Air is then bubbled into the bag before sealing. Oxygen is used in place of air if the fishes are to be transported longer than three to four hours. The bag is sealed and placed in a carton or styrofoam container. Upon arrival, the specimens should be unpackaged and acclimated to their new surroundings as soon as possible. Saltwater forms may be left in the open cartons for days and even weeks provided there is adequate areation and the temperature is properly maintained. Instant sea water mixtures may be purchased if marine specimens are to be kept on a long-term basis. Approximately twenty-five percent of the water should be changed every four weeks.

Maintenance

If a particular species of fish is to be kept active and healthy, it is necessary to become familiar with the conditions prevalent in its natural habitat. Numerous references give this information, some of which were cited earlier. Once the fishes are obtained, the most typical place to keep them is in a tank, aquarium or pool. The size of such a container, along with all the paraphernalia that belongs with it, may be purchased locally from fish centers. If fishes are to survive in captivity, then a number of factors must be considered.

Specific Maintenance Factors

CAPACITY: The number of fishes per unit space is best estimated in terms of the surface area of the tank or pool. Emmens (1962) estimated that a fish one inch in length needs 2.6 square inches of surface area. A two inch fish would require about 12.5 in^2, a four inch fish 65 in^2 and a six inch fish 180 in^2. These estimates may be lowered, i.e., more fishes may be added per unit area, if the air supply and temperature are maintained at optimum conditions. Fishes will sometimes react to an area that is too small by surfacing for air. Other fishes will not react this way but growth may become stunted if they are not removed from the crowded area.

WATER HARDNESS: Water hardness is the amount of calcium or magnesium present in the water. This is measured in parts per million (ppm). If minerals are present from 50-75 ppm, the water is regarded as *soft*. From 75-100 ppm, the water is moderately hard and above 200 ppm is considered to be very hard. Failure to replace evaporated water is the most common cause for the occurrence of hard water. This results in fishes becoming lethargic and sometimes their spawning ability is affected. Resins may be added to convert hard water into soft water. These are readily purchased in the form of water softening kits. If water conditions are naturally hard in a particular area, then it may be necessary to prepare the water that is to host the fish by adding $\frac{3}{4}$ oz. NaCl, $\frac{1}{4}$ oz. K_2SO_4 and $\frac{1}{4}$ oz. $MgSO_4$ per ten gallons of distilled water (Emmens, 1962). This usually eliminates the need for further chemical treatment.

pH: All fishes thrive only in water containing a specific amount of hydrogen ions. A pH of 6.8 to 7.2, from slightly acid to slightly basic, is in the range of neutrality and is that which is preferred for the well being of most freshwater fishes. Marine forms survive best in more basic waters, from 7.5 to 8.4. The normal tendency is for the tank water to become acid due to the buildup of carbon dioxide and metabolic waste products. The pH may be chemically regulated by adding sodium monohydrogen phosphate (Na_2HPO_4) to increase the alkalinity, or sodium dihydrogen phosphate (NaH_2PO_4) to increase the acidity. Adult fishes can tolerate extreme changes in pH much better than fry or young juveniles.

AERATION: Air is composed of approximately four parts nitrogen to one part oxygen. When air is dissolved into water the ratio becomes two parts nitrogen to one part oxygen. The amount of oxygen that is actually in the water is a function of temperature. As the temperature increases the amount of dissolved oxygen decreases. The function of aeration is primarily to remove carbon dioxide and other gases from the water and secondarily to place oxygen into the water. Both conditions occur as surface interchanges. Pumps, air stones, and filters all serve to circulate the water and enable gaseous exchanges to take place. At 75°F., the amount of dissolved oxygen should be at least five parts per million.

TEMPERATURE: The upper temperatures of waters where fishes normally live range from 50°F. to 85°F. A temperature of 72-76°F. is an average that is acceptable to most fishes and most can tolerate a gradual ten degree change without detriment. Sudden changes should be avoided since this will induce shock which may result in death. It is advisable to consult the literature to determine the temperature that is best for a given species.

CHLORINITY: Chlorine is necessary in drinking water to combat growth of harmful micro-organisms. It is, however, unfit for consumption by fish. If tap water is used for keeping fishes, it must be free of chlorine. Usually, letting the container set for at least 24 hours is sufficient to allow the chlorine to diffuse out of the water. It may be removed chemically by adding one to two grains of sodium thiosulphate per gallon of water.

FEEDING: This is perhaps the single most important factor in keeping live fishes active and vigorous. Powdered and freeze-dried foods are adequate to sustain fishes but for optimum vitality, some live food should be provided daily in the form of worms, insects and other small fishes. If, after a few hours, the food is not consumed, it should be removed from the water.

In considering all the above factors, it can be seen that one situation affects another. Temperature, for example, influences the amount of dissolved oxygen in the water and can also cause a shift in the pH. While it may not be necessary to rigidly enforce all these conditions for some fishes, the unexplained death of others can usually be traced back to the omission of at least one of them. Quite simply, fishes must be well fed in a clean and familiar environment. For the continued growth and productivity of a particular species, daily checks of the above mentioned conditions are necessary. This will pay dividends when tissue preparations are examined in hopes of finding large numbers of chromosome spreads.

Preserving Specimens

Once the organism has been sacrificed and used for research it should be preserved and deposited in a safe place. The main purpose in doing this is so the investigator, or anyone else, can later validate cytogenetic data with other morphological features. If the organisms are discarded, there is no way to associate cytogenetic findings with individual specimens. In addition to the family and correct scientific name, the organism should be labelled according to the precise locality where it was collected along with the name of the collector. If the organism was not collected, then this should be indicated. The date and the name of the cytogenetic investigator should also be given. The label should be written with waterproof ink or a soft lead pencil and placed inside the container with the specimen. The label itself should be made of acid-base resistant paper with a high rag content in order to prevent its disintegration in preservative. Small specimens can be preserved by placing them directly into a 10 percent solution of formalin. Fishes longer than several inches should be slit along the right side of the abdomen before being deposited in the preservative (Lagler *et al.*, 1962) . Fishes weighing a few pounds or more should additionally be injected with a fixative along each side of the dorsal musculature. The preserved specimen should be monitered for several weeks to insure that the label does not disintegrate or the specimen does not become severely altered. It is better to deposit the specimen in an ichthyological museum where it is given an identification number. The alternative is to store it in a safe place where it is readily available to the scientific community.

CHAPTER 2

HANDLING OF FISH CHROMOSOMES

SUPPOSEDLY, CHROMOSOMES may be obtained from any eukaryotic organism whose cells are actively dividing. Unlike other vertebrates, fishes are somewhat unique in that it is possible to get chromosomes from their tissues, suitable for analysis, in as short a period as two to three hours. There are a variety of tissues that may be selected for doing this, and since there are considerable differences in fish size and morphology, no one tissue is consistently best for this purpose. Even more variable than selecting tissues, is the choice of a method for obtaining the chromosomes. This variability is attributed to the boom in the field of mammalian cytogenetics which began in 1960. Until then, cytologists were making drawings of chromosomes that took them days or weeks to prepare and find. Suddenly, there were newer and better techniques. There was colchicine pretreatment which accumulated mataphases in large numbers; there was hypotonic solution pretreatment which caused the cells to expand and the chromosomes to separate; and, there was the air-dry method which caused the chromosomes to scatter on the slide in one focal plane. These techniques, along with many refinements and improvements, are now being used to study fish chromosomes. Before discussing these methods, however, there is the decision of what tissue to use.

SOURCES OF CHROMOSOMES

Fish chromosomes are best obtained from epithelial tissue. Since dividing epithelium lines all external and internal body surfaces, chromosomes are usually studied directly or indirectly from these areas. Figure 2-1 illustrates some tissues that are typically used for chromosome study.

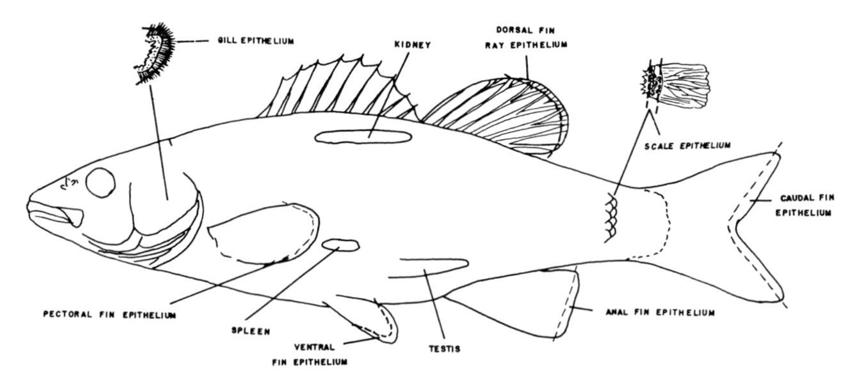

Figure 2-1. Sources of epithelial tissue in fishes from which chromosomes may be obtained.

Fins and Scales

In selecting tissues for chromosome study, it should be determined whether it is necessary to sacrifice the organism. Oftentimes the expense or scarcity of the specimen will resolve the issue. Fin or scale epithelium is well suited to studying chromosomes without sacrificing the animal. One can easily snip a marginal strip from the most active fin (usually this is the caudal or ventral fin) or pull several scales from the area of the caudal peduncle (Denton and Howell, 1969). Preparations from these tissues typically yield excellent metaphase spreads (Fig. 2-2). The selection of this tissue also has the advantage of studying chromosomes without prior treatment with a spindle inhibitor, such as colchicine. There is no danger of over contracting the chromosomes or possibly breaking chromatids.

The major disadvantage of using fins and scales is that cells which are undergoing division in these tissues are generally few in number. More figures may be obtained from regenerating fin. The margins of caudal or paired fins are snipped and after two to three days signs of regeneration may be observed. The regenerated part may then be removed for processing. If sacrificing the organism does not present a problem, then fins and scales may

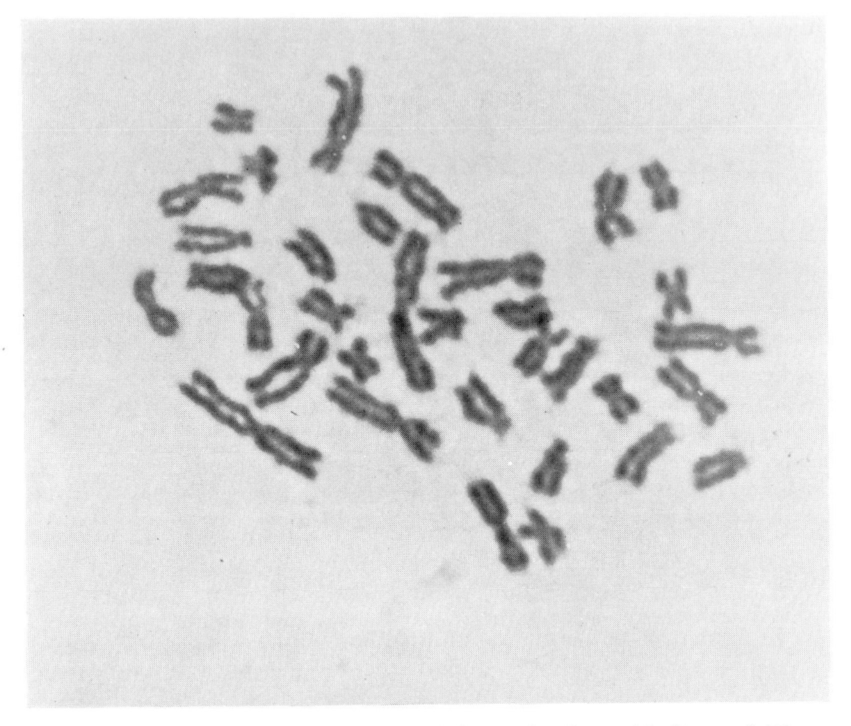

Figure 2-2. A typical metaphase spread from the fin epithelium of *Eigenmannia virescens.*

be treated with colchicine by allowing the fish to swim in a dilute solution a few hours before collecting the tissue. This treatment increases the mitotic index substantially. The most numerous figures from fin material may be obtained by using colchicine in conjunction with regenerating tissue. After signs of regeneration appear, colchicine is added five to six hours prior to processing the tissue.

Gills

The use of gill epithelium for obtaining chromosomes was first reported by McPhail and Jones (1966). Metaphase figures may be obtained from this tissue with or without colchicine pretreatment although the former is preferred for collecting large numbers of spreads. Cells from these structures are constantly in an active state of division within the different filaments. Usually, investigators work with the smaller filaments of the posterior gill arch but experience has shown that this region is not always the most active division site. The best spreads are obtained from young fishes living in fresh running waters. If the streams are polluted or stagnant then the gills become coated with mucus and debris. In preparing slides this necessitates an extra step to clean the gills, at the risk of damaging cells, before the chromosomes can be clearly enumerated. Gills, also furnish a wealth of tissue for chromosome study. The four to seven pairs of gill filaments that occur in most fishes provide enough material for preparing dozens of slides.

Spleen, Kidney and Liver

The spleen, kidney and liver have also been good tissues for obtaining chromosomes. They are generally associated with centers where blood cells are formed or destroyed. In using these tissues, colchicine treatment is an absolute necessity. This is usually accomplished by injecting a small quantity into the dorsal musculature or body cavity a few hours before the tissues are processed. Histological examination shows that mitotic figures found in the liver are derived from parenchymal cells. Spleen figures are derived from lymphoid cells, and in the kidney from myeloid hemopoietic elements.

Cornea

Corneal and conjunctival epithelium have been used by some investigators for studying fish chromosomes. These tissues are dividing most rapidly in young fishes. The percentage of dividing cells can be increased by inducing injury to the eye, or by pretreatment with colchicine. This tissue has the advantage of being readily obtainable at all times. The disadvantage is that some artistry is necessary before the technique can be used as a routine source for chromosomes.

Embryos

In some cases, embryos have been used for obtaining chromosomes. It would appear that since embryos are actively dividing that metaphase figures would be abundant in their cells. In all probability this is true, but enumerating chromosomes from these cells is another matter. Roberts (1967) discusses several disadvantages in working with fish embryos. The first problem is obtaining them and establishing identification as to species and sex. Artificial fertilization overcomes these difficulties but the procurement of mature gametes is not probable with many fishes thus eliminating the possibility of obtaining embryos. Another difficulty is the size of the embryo. The fertilized eggs of salmonids are sufficiently large so that blastomeres are easily separated from the yolk before being processed. In fishes with smaller eggs, however, the blastomeres are not easily removed from the yolk and this interferes with staining procedures. Late embryos can be separated more easily, but the number of cells undergoing division is considerably less. Colchicine will increase the number of metaphase figures but the chromosomes are reduced to pycnotic clumps which are unsuitable for analysis. As with tissues of the cornea, fish embryos can be satisfactorily used for studying chromosomes if the investigator is prepared to develop the skill to routinely remove and process the tissue.

Testes

Both mitotic and meiotic configurations are obtainable from testes materials (Fig. 2-3A-C). This is a distinct advantage in determining chromosome numbers since both diploid and haploid numbers are provided. Meiotic activity is greatest in the male shortly before the spawning season and begins to decline as the testis is filled with sperm. With some tropical forms (e. g., *Mollinesia sphenops*) figures may be obtained the year around. Colchicine injections greatly enhance the numbers of figures although a short term treatment is necessary because of the tendency for the dividing cells to become polyploid (Fig. 2-3D).

Tissue Culture

If one has the patience, time and money, then he might consider culturing tissues in order to obtain chromosomes. Mammalian tissue culture is no longer an art; fish tissue culture, to

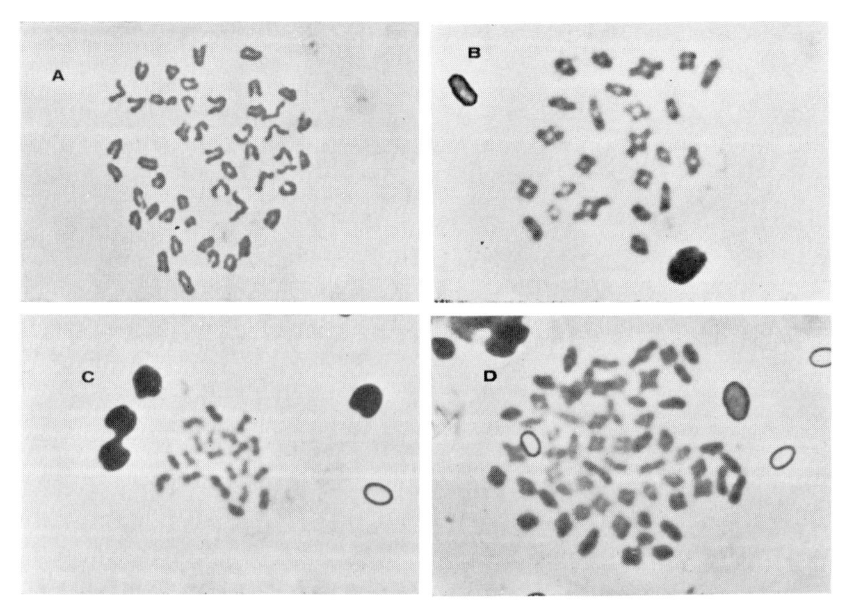

Figure 2-3. Mitotic and meiotic chromosomes from testes of *Mollienesia sphenops*. (a), mitotic metaphase containing forty-eight somatic chromosomes. (b), Metaphase I spread showing twenty-four bivalents. (c), metaphase II spread containing twenty single stranded univalents. (d), Metaphase I configuration showing colchicine induced polyploidy.

some extent, still is. Nevertheless, numerous quality metaphase figures can be obtained from actively dividing fibroblasts grown *in vitro*. Tissues are usually cultured from embryos, fins, testes, ovaries, kidneys, spleens, livers, and swim bladders. A sizable amount of tissue is needed for the sensitive digestion and centrifugation procedures that are necessary in obtaining an adequate number of cells for planting. The time involved before results can be assessed from cultures is determined by the size and age of the organism as well as the type of tissue used. This can be somewhere between six days and two months. The chromosomes obtained from these cells are of superior quality. Also the spreads are numerous. Once techniques have been standardized, this method for getting chromosomes will provide more than enough data for analyses. An excellent treatment on how to culture fish cells and tissues is given by Wolf and Quimby (1969).

Leucocytes

The most elusive technique for getting chromosomes from fishes is that of culturing blood leucocytes. This technique was first used in culturing human blood in 1960. Since then blood has successfully been cultured from other mammals as well as with birds, reptiles, and amphibians. Leucocytes normally do not undergo division in the blood stream once they are produced in hemopoietic tissue. In the presence of certain chemicals, termed mitogens, the red blood cells may be agglutinated while the leucocytes are induced to undergo a single division: These processes are routinely brought about *in vitro*. The mitogens are of two types. The one most widely used is phytohemagglutinin (PHA) which is an extract from the red kidney bean *Phaseolus vulgaris*. Phytohemagglutinin is obtainable in two forms. PHA (M) was the first to be extracted in a relatively pure state and is the one most commonly used. Normally, about 0.1 ml of this mitogen is used per 5 ml of blood. For fishes, however, this amount might need to be increased. PHA (P) is more highly purified than PHA (M) and is approximately fifty times more potent in its hemagglutinating and mitotic stimulatory capacity. For mammals, both forms of PHA have transformed lymphocytes of peripheral blood *in vitro* into blast cells which are the precursors of mitosis. Such a transformation has not been shown for fishes. The most recent mitogen is Pokeweed Mitogen (PWM) which is an extract from the pokeweed *Phytolacca americana*. This mitogen is reportedly more gentle in its blastogenic and mitogenic inducing properties than PHA in mammalian leucocyte systems. Leukagglutination is also less than with PHA. Pokeweed Mitogen has not been reported in connection with culturing fish blood. The extract from both plants are mucoprotein in nature but their specific components have not been completely defined. Their mode of action also remains a mystery. Many feel that it somehow reacts with the surfaces of monocytes and lymphocytes perhaps in an antigen-anti-body fashion.

A few investigators have been successful in culturing fish blood. Labat, *et al.* (1967) were the first to report chromosomes from cultured leucocytes using *Cyprinus carpio*. Heckman and

Brubaker (1970) reported chromosome preparations from the leucocytes of *Carassius auratus.* It was noted later by Heckman *et al.* (1971) that apparently, the choice of species had much to do with the success since the technique failed with the rainbow trout, *Salmo gairdneri.* By increasing the oxygen tension, an adequate number of spreads were produced. The common denominator in all three cases is that PHA (M) was used with a tetraploid organism. The nature of tetraploidy in the goldfish, carp, and rainbow trout will be discussed later, but in the context of culturing leucocytes, one cannot help but wonder if an increased genome in fishes is more receptive to induced mitogenesis. Attempts to culture leucocytes from diploid forms using the same methodology should settle the issue. At any rate, it does appear that some fish leucocytes are sensitive to the action of PHA.

Catton (1951) reported observing chromosomes from blood smears of the roach and trout. Recently, Barker (1972) reported a method for obtaining chromosomes from immature leucocytes in the circulating blood of the marine fish *Pleuronectes platessa.* He further reported that the technique has been successful with other marine forms. With this method, no mitogen was used and the chromosomes were harvested from the start within three hours.

It would seem that leucocyte morphology and physiology of fishes is very different from those of other vertebrates. It could turn out that one of several methods for culturing leucocytes may have to be selected to accommodate particular fish types. Since some progress has been made, however, it is hoped that a standardized method will soon emerge and the routine culturing of all types of fish blood will become a widespread reality. This method holds more promise than any other for studying somatic chromosomes of fishes because of the high quality and number of figures routinely obtainable after a reasonable period of time.

SPECIFIC CHROMOSOME TECHNIQUES

Before presenting individual techniques for obtaining chromosomes from fishes, a discussion of some basic and generalized procedures is deemed necessary.

Pretreatment With A Mitotic Inhibitor

It has already been emphasized that more metaphase figures are obtainable with colchicine treatment. This alkaloid is the most noted of all mitotic inhibitors. It was first isolated from the roots of the plant *Colchicum autumnale* in 1883 and has been widely used in plant studies. Essentially, this chemical destroys the spindle mechanism of a cell so that the chromosomes are suspended at metaphase instead of normally migrating toward the anaphase poles. In large concentrations, its effects on chromosomes can be detrimental. It can contract the chromosomes to the extent that their morphology is severely and irreversibly altered. It can also cause clumping so that individual chromosomes are not detectable. If the exposure is long it can induce polyploidy, especially in reproductive tissue. The concentrations normally used for fish tissues is from 0.01 percent to 0.1 percent for periods of one to six hours. It is thought that the colchicine wears off after twelve to fourteen hours since it is readily metabolizable by cells. Nevertheless, the organism usually dies before this occurs.

Other mitotic inhibitors consist of colcemid (deacetylmethylcolchicine), and velban (vinblastine sulfate). Colcemid is considered to be less toxic than colchicine and is used a great deal for *in vitro* studies. Velban is the most recent of the three and is also the most potent. For the same period of time, both colcemid and velban are administered at slightly lower concentrations than with colchicine.

All three inhibitors may be administered to the fish in one of three ways. For large specimens, 0.1 ml of 0.01-0.1 percent of the inhibitor per 10 grams of body weight may be slowly injected into the dorsal musculature. The same concentration may also be given intraperitoneally. The danger with this mode of application is in damaging internal organs with the needle. In instances where the inhibitor is injected, the concentrations are prepared with sterile, isotonic saline instead of distilled water. Fishes that are too small to be injected are allowed to swim for several hours in a dilute solution. Usually this is a 0.001 percent to 0.10 percent solution for three to eight hours. Because of the expense of the chemical, the container is necessarily small. This method works well, for the external tissues of fishes but the internal organs are little affected.

Hypotonic Treatment

Hypotonic treatment of fish tissue is a procedure that swells the cells and scatters the chromosomes. This is done immediately after the tissue is removed from the fish. Hypotonicity refers to an external medium that contains less dissolved material than that inside the cell. This hypotonic solution can be anything from distilled water to a solution of sodium citrate, potassium chloride or diluted isotonic saline. Treatment times normally range from seven minutes to one hour. The time will vary depending on the temperature and the consistency of the tissue used. At higher temperatures the process is speeded up and the swelling time should be decreased. Also, some tissues are more delicate than others. Testis material, for example, would require less swelling than tissue from the caudal fin. It is sometimes helpful to observe the material under a dissecting scope while it is swelling. Trial and error, however, is the only true way to arrive at a proper time for adequate spreading of the chromosomes.

Fixation

The swollen cells are chemically fixed before they are further prepared for staining. The purpose of fixation is to kill the cell without distorting its components. Of all the many fixatives, a 3:1 solution of methanol and glacial acetic acid is the one most used by cytologists. This fixative is always prepared fresh just before use. Care is exercised in handling the cells while they are swollen. They will readily burst and the chromosomes will be lost. Even after they have been fixed, the tissues will still lose cells if agitated. Cells are generally fixed for a minimum time of five minutes and often longer. After approximately thirty minutes, methanol and glacial acetic acid will begin to form methyl acetate which is a poor fixative and is undesirable for chromosome clarity. Often it is not possible to process cells immediately after fixation. In such cases the fixed cells should be placed in the refrigerator. Under these conditions the cells can be changed into fresh fixative up to twenty-four hours. This time can be extended but the risk of poor fixation becomes greater.

Staining

If chromosomes are to be viewed under a bright field microscope, then staining is necessary to produce contrast. A number of nuclear stains are cited in the literature but only three of the most widely used will be mentioned here. Aceto-orcein is by far the most commonly used. This is usually a 1 percent or a 2 percent solution in 45 percent acetic acid. The staining time under a coverslip is approximately ten minutes and in a staining dish, twenty minutes. The chromosomes appear well defined against a clear background. Synthetic orcein is commonly used but many workers prefer natural orcein. A less expensive stain is Giemsa. A 4 percent solution of stock Giemsa is prepared in a phosphate buffer at a pH of 6.4. The staining time is the same as that for aceto-orcein. Feulgen treatment of the chromosomes is the only staining process that is DNA specific. The chromosomes are first treated with Schiff's reagent. A magenta color is observed after the preparation is mounted in acetic acid. The treatment times and temperatures are critical to obtain optimum staining. Feulgen staining is time consuming and is normally used only when it is necessary to identify component nucleic acids. For routine chromosome examination aceto orcein is preferred over Giemsa or Feulgen. The chromosome outline and centromeres show up better with orcein and it is more predictable in instances where the temperature or pH may change.

Air Dried Preparations

Air drying is the newest of the methods for preparing animal chromosomes. A slurry of fixed cells is deposited on a clean slide and allowed to air dry for at least twenty minutes. Immediately after the cells are on the slide, spreading may be further facilitated by blowing straight down on the preparation or by applying warm air from a blower. Tissue clumps are quickly removed with forceps to avoid later difficulties with staining and mounting. Air dried preparations may be stored in the refrigerator in a covered container for several weeks before staining. A modification of the air drying technique is flame drying. As soon as the cells are deposited on the slide the preparation is ignited by touching the edge of the slide to a small flame. The alcohol fixative must be fresh. This adheres the cells more firmly to the slide but reduces the staining intensity of the chromosomes. It is not recommended, therefore, if the chromosomes are to be observed with bright field optics.

Temporary mounts are made by placing a drop of stain over the air dried cells and adding a coverslip. The preparation is then blotted between filter paper and ringed with Kronig's cement (Fisher Scientific). These will last for several weeks before drying out. If the slides are to be permanent then the air dried cells are stained for a required length of time in a coplin jar. For aceto-orcein this is fifteen to twenty minutes. The preparation is cleared in absolute alcohol, absolute alcohol-xylene (1:1) then pure xylene with each treatment lasting about five minutes. After drying, a drop of mounting medium is added to the slide and a coverslip applied.

Squash Preparations

The squash technique is the oldest and one of the most widely used methods for spreading and flattening metaphase chromosomes. The fixed tissue is placed in a staining solution for a suitable period of time. It is then removed from the stain and deposited on a slide with a dabbing or brushing motion. Additional stain is added if necessary before applying the coverslip. All air bubbles are removed by gently raising the coverglass with a pin point to allow their escape. A piece of absorbent paper (e.g., Whatman's filter paper) is placed over the slide. The thumb is placed over the middle of the coverslip and rolled with gentle pressure toward one end of the coverslip and then the other. Each time the stain underneath the coverslip will be absorbed by the filter paper. When the preparation becomes clear the chromosomes should be sufficiently flattened. If too much pressure is applied the stain will rush back under the coverslip creating a violent current which may break the cells.

Temporary mounts are made by ringing the coverglass with Kronig's cement. The orcein stain becomes more intense overnight. The preparation may be stored in the refrigerator for several weeks. Permanent preparations are made by placing the slide on a block of dry ice for ten to fifteen minutes. The coverglass is flipped off with a razor blade or scapel. Most of the material generally sticks to the slide. Both the slide and coverglass are treated as with air dried preparations. The slide and cover glass are mounted separately.

The advantage of the squash technique is that it is quick. If aceto-orcein is used, the swollen cells are frequently fixed and stained simultaneously since this stain contains acetic acid which is a good chromosome fixative. Many workers, however, prefer to fix the cells first in acid alcohol. The major drawbacks with the squash technique is that a large number of cells are frequently broken during the processing stages. Also, the chromosomes are often in different focal planes which creates problems when taking photomicrographs.

When the squash technique is compared with the air dried method it is felt that more figures are recoverable with the squash method and the nuclear membranes are more frequently intact which insures that all the chromosomes within the complement are present. The air dried technique, however, produces more quality figures and the chromosomes are better spread with less breakage. It is also preferred when doing DNA analyses and making autoradiographs. In essence, the investigator should try both methods and decide which one best meets his need.

CHROMOSOMES FROM SCALE AND FIN EPITHELIUM
(Denton and Howell, 1969)

1. Place a specimen that is six inches in length or shorter into a 0.01 percent aerated colchicine solution for four to five hours. Remove and wash under running water.*
2. Scrape a number of scales from the area of the caudal peduncle and place in a syracuse dish containing distilled water.
3. Snip the margins of the dorsal, caudal and paired fins and place in distilled water separately from the scales. Allow to swell for approximately one hour.

*This step is omitted if the organism is not to be sacrificed.

4. Carefully aspirate the water from around the scales and fins so that a minimum of movement is produced.
5. Slowly add freshly prepared methanol-glacial acetic acid (3:1) down the side of the dish.
6. Aspirate the fixative water mixture and add fresh fixative for five to ten minutes.
7. To air dry, remove the tissue with forceps and dab on a clean slide. Remove large tissue clumps and allow to air dry. Stain under a coverglass with 2 percent aceto-orcein for ten minutes. Blot between filter paper and ring with Kronig's cement (Fisher Scientific).
8. To squash, remove the fixative in step 6 and stain in orcein for ten minutes. With forceps, brush cells from the edge of the tissue along the surface of a clean slide. Add a drop of stain to the slurry of cells, apply a coverslip and place between filter paper. Squash until the preparation becomes clear. Ring with Kronig's cement. Permanent preparations may be made by the dry ice technique of Conger and Fairchild (1953).

Comments: Two to three days before the scale and fin epithelium are to be processed, trim the edges of all fins so that regeneration may occur. The new growth is lighter in color and is readily detectable with the naked eye. The following causes may be responsible for poor preparations: dirty background: The cells were ruptured while swelling or excess material was not pressed from under the coverslip. Chromosomes clumped together: The swelling time was not long enough. Cells clumped together: Fixation was poor. Preparation granular: The stain needs filtering. Cells too light or too dark: The preparation was either understained or overstained.

The cause for cell breakage using the squash technique may be due to the suspension of cells in the hypotonic solution or fixative too violently, or by squashing too hard.

OBTAINING CHROMOSOMES FROM GILL EPITHELIUM
(McPhail and Jones, 1966)

1. Inject 0.1 ml of 0.01-0.05 percent (higher levels for larger fishes) colchicine into the anterior-dorsal musculature.
2. Place in a well aerated container for one to two hours.
3. Pith the specimen and remove the posterior gill arch.
4. Place in distilled water for thirty minutes.
5. Stain in aceto-orcein for fifteen minutes.
6. Deposit a slurry of cells on a clean slide, squash, and ring with a mixture of lanolin-paraffin.
7. Air dried preparations may be made with the following modifications: Place in distilled water for one hour; fix in acid alcohol for ten minutes; air dry on a clean slide and stain with orcein as with scales and fins.

Comments: This technique will produce numerous cells but the mitotic index will vary with each fish. The greatest number of figures are obtained from young active specimens. For fishes that are too small to inject, comparable results may be obtained by allowing the organism to swim around in 0.01 percent colchicine for four to six hours. Marine fishes are injected with 0.5 percent colchicine for twelve hours and swollen in 50 percent sea water for one hour.

CHROMOSOMES FROM GILL EPITHELIUM
(Lieppman and Hubbs, 1969)

1. Inject 0.03-04 ml of 0.025 percent colchicine into the right epaxial muscle posterior to the right gill arch and operculum. Maintain for three hours at room temperature before sacrifice.
2. Remove the forward three gill arches on the right side of the animal and incubate in a hypotonic solution. For freshwater specimens taken where little or no saltwater intrudes, incubate in double-distilled water for thirty minutes at room temperature. For freshwater-brackish water specimens, incubate in 1 percent solution of TM (Maio and Schildkraut, 1966) for twenty minutes.

3. Fix gill arches in 50 percent acetic acid for twenty minutes.
4. Air dry gill arches for one minute after removal from the acid.
5. Transfer to coverslips, add one drop of 1 percent aceto orcein, and cover with a watch-glass for ten minutes.
6. With jeweler's forceps, gently agitate gill arch over the coverslip, which sheds epithelial cells. Remove clumps.
7. Place a slide over the coverslip and turn right side up. Place between filter paper and apply thumb pressure from the center of the coverslip until the thumb is off the coverslip. Repeat with increasing pressure. If the chromosomes are not sufficiently spread, heat by passing over an alcohol lamp that is protected by filter paper and repress with the thumb.
8. Seal with Kronig cement and store in the refrigerator.

Comments: This technique was originally done on cyprinid fishes weighing one to two grams. Failure to aerate the water during colchicine treatment resulted in a lowered mitotic index. Distilled water treatment of gills from a brackish water specimen resulted in exploded cells. Since only the right gill arches are removed, the left side is left unmarked for identification purposes. Care in handling of swollen or fixed gill arches reduces loss of cells due to agitation. Step 5 dissociates the cells of the gill arches while leaving the cellular membranes intact.

✳ CHROMOSOMES FROM SPLEEN, KIDNEY AND GONAD

Ohno (1965), Becak (1966)

1. Two hours prior to sacrifice, inject 0.5-1.0 ml of 0.5 percent colchicine into the dorsal musculature of the specimen.
2. Remove spleen, kidney and gonad and mince into approximately 2mm cubes.
3. Swell in cold distilled water at a pH of 7.0 for fifteen minutes.
4. Fix in 50 percent acetic acid for fifteen to thirty minutes and squash.
5. Remove the coverglass by the dry ice method and hydrolyze

the tissue in 1 N HCl for ten to fifteen minutes at 60°C before staining with Giemsa.

Comments: This technique may be modified for air drying by swelling the cubes for forty minutes to one hour in room temperature distilled water, fixing in acid alcohol for ten minutes, air drying, and staining in 2 percent orcein for ten minutes. Large amounts of tissue are preferable when using either method.

Kidney cells, as well as testis cells, may also be collected according to the method of Ojima and Hitotsumachi (1967). The tissues are minced with scissors in a balanced salt solution and the dispersed cells collected in a centrifuge tube, treated with hypotonic solution, and fixed with acetic alcohol. The chromosomes are spread by the air drying method. This technique provides quality spreads with a minimum of cell loss.

CHROMOSOMES FROM EMBYRO SQUASHES OF SALMON
(Simon, 1963, 1964)

1. Select fertilized eggs that have incubated for forty-eight hours at 50°F.
2. Fix in glacial acetic acid and isopropyl alcohol (1:99).
3. Separate germinal cells (blastodiscs) from the yolk by peeling away the chorion and picking the exposed disc of cells from the yolk surface with curved forceps.
4. Stain blastodiscs in approximately 1 ml of 2 percent aceto-orceinpropionic acid (19:1).
5. Transfer stained material to clean slide, cover with drop of stain, squash, and ring coverglass.
6. Allow at least two days before examining.

Comments: Fertilized eggs that have incubated for 48 hours at 50°F contain from 15-50 percent cells in metaphase. Beyond this time, the chromosome size diminishes as the egg reaches gastrulation. Fixation and fat extraction may be facilitated before staining and squashing by treating the blastodisc in acetone for five minutes and rinsing briefly in two changes of acetic-alcohol.

CHROMOSOMES FROM PRIMARY TISSUE CULTURES
(Roberts, 1964, 1970)

Equipment

Alcohol, cotton

Cold plate with magnetic stirrer or container with stirrer assembly and crushed ice.

Transfer hood

Small bunsen burner or alcohol lamp

Clinical centrifuge

Slides, coverslips

Rubber bulbs

Aceto-orcein, natural, 1 percent (George Gurr, Ltd)

Cold room or incubator

Kronig's cement

The following items should be sterilized by autoclaving.

Some may be purchased in the sterile, disposable form (Appendix).

Surgical scissors, assorted sizes

Forceps, assorted sizes

Petri dishes

Centrifuge tubes, 12 ml

Pasteur pipettes

Leighton tubes (16 × 85 mm) containing coverglasses (10.5 × 50 mm)

Silicone rubber stoppers

Filter assembly for sterilizing liquids

Fluted flask (50 ml) with teflon stirring bar and chrome cap.

The following reagents should be bought in sterile form or sterilized by filtering through a Millipore filter with a porosity of 0.2 microns.

Hanks Balanced Salt Solution (BSS)

Trypsin, 2.5 percent

Fetal calf serum

Penicillin-Streptomycin solution
Minimal Essential Growth Medium (MEM)
Sodium citrate, 0.9 percent
Colcemid (10 mcg/ml) in Hanks Balanced Salt Solution

Procedure

1. Select a specimen at least eight inches in length, preferably a male. Sacrifice by severing the spinal cord immediately behind the head. Rinse in tap water and swab with 70 percent alcohol.
2. Collect 2-4 cm^3 of tissue (testis, spleen, kidney, swim bladder, ovary, or liver) as aseptically as possible. Instruments may be sterilized by dipping in 70 percent alcohol and flaming before each use.
3. Transfer tissue to a chilled petri dish containing Hanks Balanced Salt Solution.
4. Mince into small pieces and wash three times with cold saline.
5. Transfer minced tissue to a fluted flask containing freshly prepared digestion medium:
   17.0 ml Hanks Balanced Salt Solution
   2.0 ml of 2.5 percent Trypsin solution
   1,000-4,000 units penicillin and 1,000-4,000 mcg/ml streptomycin (depending on risk of contamination)
   0.5 ml Fetal Calf Serum

   Sterilize if necessary
6. Place digestion flask with tissue on a magnetic stirrer at 4°C for one hour.
7. Discard supernatant and replace with an equal volume of freshly made digestion mixture. Allow digestion to occur for one and one-half to three hours. Transfer supernatant to sterile centrifuge tubes.
8. Sediment by cold centrifugation at 200 g for ten minutes. Button should contain 0.1 ml or more of cells. Discard the supernatant.
9. Rinse in fetal calf serum to neutralize the trypsin.
10. Resuspend cells in the following growth medium by trituration:

800 ml MEM (Eagle)
56 ml fetal calf serum
8 ml L-glutamine
100 units/ml penicillin and 100 mcg/ml streptomycin

Adjust pH to 7.2 and sterilize if necessary. The medium should be stored in the frozen state.

11. Inoculate Leighton tube containing coverglass with 1.5 ml of cell suspension. Plug with silicone rubber stopper.
12. Incubate for five to six days at 19 ± 1°C.
13. Four to six hours before harvesting cells, add 1.5 ml colcemid per tube.
14. Discard medium and colcemid and replace with 3 ml of 0.9 percent sodium citrate for ten minutes.
15. Remove citrate and stain with 1 percent aceto-orcein for five minutes.
16. Invert coverglass over slide containing a small drop of stain. Remove air bubbles and squash lightly, Ring with Kronig's cement or a 50 percent paraffin-50 percent lanolin mixture with a small piece of wax pencil for color. Slides may be made permanent by the dry ice method.

Comments: The cell suspension from the digestion mixture should be enough to seed five to ten tubes. By way of personal communication, the author reports that postspawned ovary should be the tissue of choice. Other tissue should be used only when the ovary cannot be obtained.

The critical steps include destroying the cells with trypsin during the digestive procedure, contamination during handling, and changes in temperature and pH during the incubation period. Staining a sample of material from the digestion mixture with trypan blue will determine cell viability. Contamination can be controlled to a minimum by sterilizing all materials and using enough antibiotics to restrict further growth of bacteria. The silicone rubber stoppers will prevent loss of CO_2 and stabilize the pH. It may be necessary to gas with 5 percent CO_2 before incubating in order to start at the proper pH level. The temperature can be adequately maintained during incubation with a dependable incubator or cold room and daily vigilance.

There is some variability in the length of time before the fibroblasts reach the log phase of growth. If embryonic cells are used, it is possible to process them for chromosomes within 48 hours (Rachlin *et. al.,* 1967). If older tissues are used, the cells will take longer to reach the exponential growth phase. This could be from one to three weeks.

The above squash procedure may be modified for air drying by continuing from step 13. The medium and colcemid are replaced with 3 ml of 0.075 M KCl for seven minutes. The KCl is removed and fixed in two changes of methanol-acetic acid (3:1) for ten minutes each. The coverslips are air dried for at least thirty minutes and stained in 2 percent aceto-orcein by inverting over a drop of the stain for ten minutes before blotting and ringing with permount or Kronig's cement.

In the event that cells are seeded in larger vessels, such as falcon plastic or prescription bottles, then they may be removed by treatment with cold 0.025 percent trypsin for thirty seconds and immediately neutralized with fetal calf serum. The cells are collected by centrifugation, washed with growth medium, and swelled with 0.075 M KCl for seven minutes. The KCl is replaced with methanol-acetic acid for fifteen minutes in the refrigerator. The fixative is changed twice. After the second change, all the fixative but 0.5 ml is drawn off and the cells resuspended. One drop is deposited on a clean slide and air dried before staining.

CHROMOSOMES FROM CULTURED OVARIAN CELLS
(Chen, 1970)

This method was reported for two species of *Fundulus* which are found in salt, brackish, and fresh waters. Both specimens were collected from salt water and kept in aquaria for one to three months before culturing.

1. Females are removed from the saltwater aquaria, rinsed in tap water, and sacrificed by severing the spinal cord. The outer surface is sterilized with absolute ethanol. Instruments are sterilized by alcohol dipping and flaming before each use.
2. Remove the ovaries and place in 5 ml of 0.25 percent viokase (Gibco) and mince. Transfer the fragments to 10 ml of fresh viokase at room temperature on a magnetic stirrer for fifteen minutes. The trypsinization is stopped by adding 1 ml of Fetal Bovine Serum (FBS) .
3. Centrifuge, collect, and resuspend cells in 1 ml of F-10 or F-12 culture medium and inoculate into 30 ml Falcon culture flasks containing 4 ml of medium. Gas with a mixture of 5 percent CO_2-95 percent air.
4. Incubate at 31°C for five to seven days with one feeding after three to four days by replacing 2 ml of old medium with fresh medium.
5. Six to eight hours before harvesting, add colcemid to give a final concentration of 1 mcg/ml.
6. Decant the medium and add 3 ml of hypotonic trypsin-versene solution for ten minutes. Collect the suspension of cells in a centrifuge tube prefilled with 0.3 ml of FBS. Clumps or attached cells are dispersed by gentle trituration.
7. Centrifuge at 1000 rpm for ten minutes, decant to leave 0.5 ml. Resuspend and hypotonize with 4 ml of 0.075 M KCl for 10 minutes.
8. Again centrifuge and decant leaving 0.5 ml and resuspend pellet. Fix in acetic-alcohol (1:3) by slowly adding up to 4-5 ml for fifteen minutes.
9. Repeat step 8 and finally resuspend pellet. Add about three drops of suspension to a chilled wet slide and air dry. Leave overnight before staining.
10. Immerse air dried slide into freshly filtered aceto-orcein (1.5 g of Gurr's synthetic orcein in 100 ml of 65 percent acetic acid) for fifteen minutes. Remove excess stain in absolute alcohol and mount under a #1 coverslip in Canada balsam.

Comments: The author reported similar karyotypes with both the squash and air dried treatments but favored air drying because of greater clarity and definition.

CHROMOSOMES FROM THE LEUCOCYTES OF CYPRINUS CARPIO

(Labat *et al.*, 1967)

1. Obtain 0.5 ml of blood by heart puncture with a heparinized syringe. Place the blood immediately with 20 ml of the following culture media:
   - 6 ml of whole egg ultrafiltrate
   - 30 ml of 2 percent lactalbumin hydrolysate
   - 11.8 ml of fetal calf serum
   - 6 ml of Phytohemagglutinin M
   - 25,000 units of penicillin G
   - 5 ml of streptomycin
2. The cells are cultured in pyrex tubes at 20°C for seventy-two hours. The cultures are agitated every twenty-four hours.
3. Colchicine is added a few hours before the cells are to be harvested.
4. The cells are harvested and air dried by conventional methods.

Comments: The author reports that chromosomes obtained in this way are suitably dispersed and their morphology is clearly defined.

CHROMOSOMES FROM THE LEUCOCYTES OF TROUT GROWN IN VITRO

(Heckman *et al.*, 1970, 1971)

1. Withdraw 4-8 ml of blood by cardiac puncture from the heart of a large specimen with a 10 ml syringe containing 0.5 ml of anticoagulant (Difco).
2. Sediment the erythrocytes by centrifuging 100 g at 9°C.
3. Place 4 ml of the following medium into culture bottles 2 cm in diameter and 7 cm deep:
   - 100 ml MEM (Eagle spinner modified, Difco)
   - 10 ml Fetal calf serum (Gibco)
   - 0.2 ml L-glutamine (200 mole/liter, Gibco)

4. Saturate with pure oxygen and add 0.4 ml of leucocyte rich plasma and 0.3 ml of phytohemagglutinin M (Difco).
5. Saturate the atmosphere of the container with pure oxygen and seal.
6. Incubate for 120 hours at 19°C at a pH of 7.0.
7. Four hours before harvesting the cultures, add colchicine mixed in Hanks balanced salt solution to give a final concentration of 0.005 mg/ml.
8. If there is a carpet of agglutinated red blood cells, it should be removed with a Pasteur pipette by dipping into and not of the medium to wash off any leukocytes clinging to it.
9. Transfer contents to a 12 ml centrifuge and spin for five minutes at 800 rpm. Remove all but 0.25 ml of medium.
10. Add 5-6 ml of hypotonic solution (0.067 g $CaCl_2$ and 0.2 g sodium citrate in 100 ml water) for twenty minutes at 22°C.
11. Centrifuge at 750 rpm for six minutes, discard all but 0.25 ml of supernatant, and resuspend.
12. Add 4 ml of freshly prepared methanol-glacial acetic acid (3:1) by adding 0.5 ml, allowing to equilibrate, then adding the remainder. Allow the cells to fix for fifteen minutes at room temperature.
13. Repeat steps 11 and 12.
14. After the second fixation procedure, centrifuge at 750 rpm, remove all but 0.5 ml fixative, and resuspend cells.
15. Air dry on clean slides and stain with a mixture of one part of saturated alcoholic crystal violet and four parts of 1 percent ammonium oxalate.

Comments: The authors reported from 10-1000 spreads per slide when oxygen was used, but rarely up to ten figures without oxygen treatment.

CHROMOSOMES FROM THE LEUCOCYTES OF SOME MARINE FISHES

(Barker, 1972)

1. Draw 1 ml of blood from the heart of a large marine specimen with a sterile heparinized syringe and transfer to a centrifuge tube.

2. Spin at 90 g for three minutes, withdraw leucocyte rich plasma and deposit into a glass bottle containing 7 ml of the following culture medium:
   5 ml Medium 199 (adjusted to osmolarity with NaCl)
   2 ml Fetal calf serum
3. Add 0.1 ml 10^{-3} M Colcemid (Ciba) and allow to incubate for two hours at 10°C.
4. Centrifuge at 225 g for five minutes, discard supernatant.
5. Swell in 5 ml of 1 percent sodium citrate for fifteen minutes at room temperature.
6. Centrifuge at 225 g for five minutes, discard supernatant.
7. Add 5 ml of acetic-alcohol (3:1) with gentle agitation, repeat step 6.
8. Suspend cells in fresh fixative for thirty minutes at 0°C.
9. Repeat step 6 and add 5 ml fresh fixative.
10. Repeat the last step with 50 percent glacial acetic acid.
11. Centrifuge at 225 g for five minutes. Remove the supernatant and make a final dilution with 1-2 ml 50 percent glacial acetic acid.
12. Dispense four drops of suspension per slide and hold on a hot plate at 90°C until dry. Stain overnight in 2 percent lacto-acetic orcein, dehydrate and mount in Euperal.

Comments: The author reports that it is important to emphasize the use of spleen tissue for chromosome display, as occasionally the number of metaphase cells in the circulating blood of an individual fish can be quite low (personal communication). This is accomplished by macerating spleen tissue in 2 ml of Medium 199, drawing off the cell suspension and making up to volume as described in step 2. The leucocyte technique has been successfully used with the plaice *(Pleuronectes platessa)*, the dab *(Limanda limanda)*, the thornback ray *(Raia clavata)*, the eel *(Anguilla anguilla)* and the sole *(Solea solea)*.

PERMANENT DRY MOUNTED CHROMOSOMES FROM TELEOST FISH

(Stewart and Levin, 1968)

This technique was used in preparing chromosomes from branchial epithelium, kidney cells, testes, and blastodiscs. Modifications necessary for the preparation of chromosomes from these tissues are noted where applicable for each step.

1. Inject 0.01-0.02 ml of 0.01 percent colchicine intra-muscularly for two to six hours at 16-20°C under heavy aeration. In the case of embryos, no colchicine is used. Incubate fertilized eggs for thirty-six hours at 18°C.
2. Sacrifice the fish by pithing and remove the fourth branchial gill arch, testes, and kidney. Remove all mucous, blood, and other debris adhering to the gill lamelliae. Mince testis and kidney into 1-2 pieces. Remove the blastodiscs from the eggs, separate from the yolk material, and place in ringers solution.
3. After cleaning, transfer the gill arch, minced testis, or kidney to 0.1 M KCN for twenty to ninety seconds, depending on size.
4. Transfer to triple glass-distilled water for three to five minutes or until visibly swollen. Place blastodiscs in distilled water for ten minutes.
5. Fix all tissue in 45 percent acetic acid for fifteen minutes except blastodiscs.
6. Apply tissue to clean slides with a *painting* motion or disperse with pressure from a clean scapel blade and allow the smear of monolayered cells to air dry. For blastodisc cells, fixation and smearing is done simultaneously in a drop of fixative on a clean slide.
7. Stain for one hour in 5 percent Giemsa, rinse in distilled water, air dry, and mount with Permount.

Comments: The KCN treatment destroys the osmoregulatory mechanism of the cells and subsequently increases the inflow of hypotonic solution. Slides show no deterioration after twelve months of storage. The mechanical disruption involved in squashing is eliminated. It is not necessary to precede the dialysis of blastodisc cells with KCN treatment.

CHROMOSOMES FROM EMBRYOS OF THE ZEBRA FISH

(Endo, 1968)

1. One day old eggs are collected and placed on a petri dish in an isotonic salt solution where the membranes are removed using forceps and knife.
2. The embryos are transferred to a watch-glass containing a Pasteur pipetful of isotonic solution with 0.2 micro-gm/ml of Colcemid, and minced thoroughly.
3. The cell suspension is transferred to a 10 ml beaker along with two to three more Pasteur pipetfuls of isotonic solution with Colcemid, then placed on a magnetic stirrer for twenty minutes.
4. After centrifuging at 600 rpm for five minutes, the supernatant is decanted and 2-3 ml of hypotonic solution (calf serum: water, 1:5) is added and kept at room temperature for twenty minutes.
5. The cell suspension is centrifuged and fixed in 2-3 ml of fresh fixative (acetic acid: methanol, 1:3) without disturbing the cells.
6. The fixative is changed two to three times before preparing air dried slides as with leucocyte cultures.

Comments: This technique takes advantage of the fact that young embryos are rapidly dividing and by treatment with colchicine, metaphases can be collected with steps similar to those for harvesting leucocytes or fibroblasts. Embryos younger than one day or older than thirty-six hours do not yield sufficient numbers of metaphase cells.

CHROMOSOMES FROM FISHES WITH CALCIUM PRETREATMENT

(Subrahmanyam, 1969)

1. Inject 0.1 percent solution of calcium chloride intraperitoneally with the following dosage according to the size of the fish:

5-10 cm - - - - - - 0.50 ml
10-15 cm - - - - - - 0.75 ml
15-20 cm - - - - - - 1.00 ml

2. Three hours after the injection of calcium chloride, inject intramuscularly 0.2 percent colchicine solution in distilled water (1.00 ml/50 g body weight).
3. Keep in well aerated aquarium for twelve to sixteen hours before sacrificing.
4. Remove gill arches, kidneys, and gonads.
5. Dialyze gills in glass-distilled water and kidneys and gonads in 0.9 percent sodium citrate for one to fifteen minutes.
6. Stain in 2 percent aceto-orcein and squash.

Comments: The above technique was used on the estuarine mud skipper *Boleophthalmus boddaerti.* Treatment with $CaCl_2$ is believed to enhance mitotic divisions and to counteract the contraction of chromosomes normally produced by colchicine.

CHROMOSOMES FROM MARINE GRUNT FIN IN TISSUE CULTURE

(Clem, 1961, Regan, 1968)

1. Select a mature specimen and wash thoroughly in tap water. Remove fin tissue and wash three times for ten minutes each in Hanks balanced salt solution containing penicillin (300 units/ml), streptomycin (150 mg/ml), and mycostatin (300 units/ml).
2. Mince the washed tissue into small fragments and add to ten volumes of 0.25 percent trypsin prepared in Mg free Hanks BSS containing penicillin (100 units/ml) and streptomycin (50 mg/ml). The Hanks solution contains 0.206 M NaCl instead of the usual 0.137 M used with mammalian systems.
3. Allow the trypsin to act on the tissue fragments in an Erlenmeyer flask over a magnetic stirrer at room temperature for thirty minutes. Discard and add fresh trypsin for one to one and one-half hours.
4. Neutralize trypsin with one volume of human serum and

collect cells by centrifuging at 700 rpm for seven minutes at 5°C.

5. Discard supernatant and resuspend cells in growth medium consisting of Eagles Basal Medium, (0.206 M NaCl), 10 percent calf serum, 100 units/ml of penicillin, 50 mcg streptomycin/ml and 100 units of mycostatin/ml.
6. Seed Leighton tubes containing coverglasses with 1 ml of cell suspension containing approximately 8×10^5 cells.
7. Incubate at 20°C for ten days to two weeks, changing the medium at four day intervals.
8. Chromosome preparations are made by adding colchicine (2×10^{-5}M) twelve to twenty-four hours before collecting cells.
9. Remove the medium and add 0.05 percent trypsin in calcium and magnesium free balanced salt solution for ten minutes. Centrifuge, wash once in Hanks BSS and resuspend in 0.05 percent sodium citrate at room temperature for fifteen minutes.
10. Obtain pellet of cells through centrifugation and fix with 3:1 methanol-glacial acetic acid for two hours. Transfer to fresh fixative, spread on slides, and stain in 2 percent aceto-orcein.

CHROMOSOMES FROM CORNEAL AND CONJUNCTIVAL EPITHELIUM

(Drewry, 1957)

1. The specimen is sacrificed and the eyes carefully removed to a hypotonic solution consisting of Holtfretter's solution with one-half the normal concentration of NaCl and KCl, for twenty to forty minutes.
2. The epithelium is fixed by dipping into Carnoy's fixative (three parts ethanol, one part glacial acetic acid). This facilitates its removal from the cornea.
3. Small patches of epithelial tissue about 1 mm^2 are placed in drops of stain-fixative (2 percent orcein in one part glacial acetic acid, 1 part lactic acid, one part H_20) on microscope slides and covered with glass cover slips.
4. The cells are separated by sharp taps above the pieces of tissue with head of a steel insect pin then squashed by working

the fluid to the edge of the glass with light but firm strokes of the pinhead.

Comments: If the quantity of fluid is correct, the cells are kept flattened by capillarity between the slide and coverglass. Chromosomes spread in this way are reportedly better than that obtained with a mechanical press and only a few cells are disrupted during squashing. The technique fails to work with stain fixatives which do not contain lactic acid.

MEIOTIC CHROMOSOMES FROM TESTIS MATERIAL (Denton)

1. Depending on the size of the specimen, slowly inject 0.1-1 ml of sterile, isotonic colchicine (0.1 percent) or colcemid (10 mcg/ml) into the intraperitoneal cavity two to four hours before sacrificing (care should be taken not to damage internal organs) .
2. Remove the testis, wash in tap water, and transfer to freshly distilled water for twenty to thirty minutes. If the material is to be squashed, swell in 0.9 percent sodium citrate for ten minutes.
3. For air dried preparations, draw off the hypotonic solution and fix in 3:1 methanol-glacial acetic acid (add slowly) for ten minutes at room temperature. Change the fixative a second time for five minutes. With curved tipped forceps, apply the tissue to a clean dry slide. Quickly remove tissue clumps and allow to air dry for at least thirty minutes. Stain for ten minutes in 2 percent aceto-orcein under a coverglass, blot between filter paper, and ring with Kronig's cement.
4. For squash preparations, immerse the swollen tissue directly into 2 percent aceto-orcein for ten to twenty minutes. Take precautions to prevent the stain from evaporating while staining. Remove the tissue with forceps and apply a uniform slurry of cells to a clean slide, add a drop of fresh stain, apply a coverslip, remove air bubbles, and squash between filter paper using thumb pressure. Ring with Kronig's cement.

Comments: Testes material is very sensitive to colchicine or colcemid. With exposures longer than four hours, polyploid figures become numerous and obscure the true haploid number. For some fishes the testes are highly active the year around. For others, they are most active one to two months preceeding the spawning season. During this time, colchicine treatment is usually unnecessary.

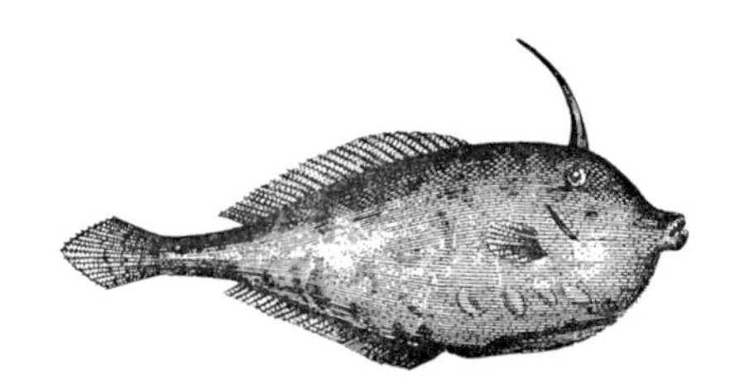

CHAPTER 3

EXAMINING CHROMOSOMES

IN THE PREVIOUS CHAPTER specific methods were given for preparing chromosomes from various fish tissues. Once chromosomes are obtained, their number and structure must be examined, evaluated and recorded. This chapter will deal with some of the routine and special instrumentations and methods for working with chromosome preparations so that useful and meaningful data may be collected and applied to specific problems and situations.

LIGHT MICROSCOPY

One very important difference between fish chromosomes and those of other vertebrates, is that they are generally smaller and more difficult to enumerate. The cytological methods presented previously require an instrument that reveals details which are less than a micron in size. Furthermore, it is required that the instrument be used competently. This instrument is the compound light microscope. Twenty years ago microscopes were all very similar, and they were also inadequate for studying fish chromosomes. Today's microscopes reflect tremendous advances and there are many models on the market that will resolve these chromosomes in detail. The problem is to determine which one is the best investment that will meet the need. On the one hand there is the student microscope, and on the other, the research microscope. Student microscopes share the same inadequacies as the older models; they are unsuitable for the detailed study of fish chromosomes. It therefore becomes imperative to purchase one of research quality. Sometimes the distinction between the two microscopes is not clear. The most dependable guideline, perhaps, is price. The optical quality is usually better with the more expensive models. Numerous makes of research microscopes are available, and since several hours a day may be spent looking through one, it is well worth the time and money to survey the market.

Magnification and Resolution

The magnification of a microscope is measured as the product between the magnification of the objective and that of the eyepiece or ocular. The eyepiece generally has a magnification of 10X while the objective may have different magnification values The most common ones are 10X, 40X and 100X. Sometimes there is a tube factor between the objective and eyepiece that increases the overall magnification by a factor of 1.5 to 3. This means that the typical range of magnification for the compound microscope is from 100X to 3000X. Many inexpensive microscopes will magnify up to 2000X, but most of this enlargement is said to be *empty* or spurious magnification. Even with research quality optics, the most useful magnification at the upper limit is approximately 1250X. Little purpose is served therefore by using eyepieces at more than 10X-15X.

An optical property that is just as important as magnification is resolution. This is the ability to resolve or separate two points that lie close together. It is this resolving power that sets the limits of useful magnification. This property is built into the lens system and is a function of the numerical aperature (N.A.). The numerical aperature is found as $n \sin u$, where n is the refractive index of the media between the lenses, and u is the greatest angle of light to the optical axis of the objective. In addition to the numerical aperature, resolution is dependent upon the wavelength of light within the optical system. Normally, visible light is used which contains several separate wavelengths. With these two properties, the resolution of a compound microscope can be estimated from the following relationship:

$$\text{Resolution} = \frac{\text{Wavelength of Light Source}}{\text{N.A. of Objective} + \text{N.A. of Condenser}}$$

The optimum numerical aperature for the oil immersion objective is 1.4. Sometimes the optimum value for the condenser is also 1.4 but is usually less. The midpoint value for the wavelength of light in the visible spectrum may be approximated at 0.55 microns. When these values are placed in the above formula, the resolution is calculated to be about 0.2 microns, which is the limit of resolution for the light microscope. This means that potentially, the oil immersion system of the light microscope can resolve points in an object that are separated by a minimum distance of about 0.2 microns. The numerical aperature, as well as the magnification value, is inscribed on both the objectives and condensers of better microscopes.

Objectives and Eyepieces

Objectives are the most critical and expensive of all the components of the microscope. The performance of the entire system depends on the quality of the lenses in this structure. Different materials go into the complicated making of these lenses and there may be as many as eight or more in one objective. There are four principal types of objectives. These are the achromats, fluorites, apochromats, and the planapochromats. The lenses in these objectives are designed to overcome two types of optical aberrations, viz. spherical and chromatic. Spherical aberrations are discrepancies between the light path through the center and periphery of the lens system. Chromatic aberration is the focusing of different colors of a light beam at different points along its path.

Achromatic objectives are the types most commonly used for routine work and they are the least expensive. The chromatic aberration in these objectives is corrected for two colors and the spherical aberration for one color. Fluorite objectives compensate for the same aberrations as the achromats, but does so more effectively. Apochromats correct for three spherical and two chromatic colors respectively, but there is usually a difference between the focus in the center and edges of the field. Planapochromatic objectives correct for this difference and provide a totally flat field with everything in one focal plane.

Objectives are either dry or oil immersion. The dry objectives are corrected for working with specimens that vary in thickness and refractivity. With high power objectives oil should be used between the coverslip and objective to increase and unify the refractive index. Immersion oils should be bought from the manufacturer of the microscope. When different oils are used, this will sometimes result in an opaque mixture that completely obliterates the image. Microscope slides and coverglasses of standard thickness are adequate for use with dry objectives. Slides of standard thickness, or less, and size 0 coverglasses, are recommended for use with oil immersion.

The eyepiece is used in conjunction with the objectives to view the real or true image which is located in the microscope tube just below the ocular. The eyepiece can be one of two types, Huygenian and compensating. Huygenian eyepieces are made of simple lenses and is compatible with achromatic objectives. Compensating eyepieces are required for use with the fluorite, apochromatic, and planoapochromatic objecties in order to correct for the different colors emanating from the more critically corrected lenses An ocular with a magnification of 10X is generally best for use with high power objectives. Magnifications above this will result in a distortion of the image and spurious enlargement.

Condensers and Light Sources

The sub-stage condenser forms the third lens system in the microscope and is used to focus the column of light from its source onto the specimen thereby filling the aperture of the objective with uniform light. The Abbe condenser is the simplest and most common type found on less expensive microscopes. These are two other types that are better for work requiring higher resolution. The aplanatic type condenser gives a reasonably good image when the light source is of one color. The aplanatic-chromatic or chromatic condenser is the highest grade lens system and approaches the quality of those found in the objectives. These condensers prevent the escape of light rays that normally produce glare. This does not produce better resolution but it does give better contrast. To obtain optimum performance with condensers of high N. A., immersion oil should be placed between the slide and the top lens. This reduces the light reflection and enables the lens system of the condenser to perform the function for which it was constructed. For routine use this is usually regarded an unnecessary trouble and is seldom done. The aplanatic or achromatic condenser should be used in conjunction with fluorite or apochromatic objectives. Ideally, the N. A. of the condenser should be the same as, or only slightly less than, the N. A. of the oil immersion objective.

The light source for the research microscope is often minimized when selecting an optical system. This is unfortunate because, without a good light source, all the lens systems of the microscope could not fulfill their designed functions. Preferrably, the light source should be built into the microscope. It should consist of artificial light from a 6 volt bulb in which the intensity can be regulated by a variable transformer. It should also have a field diaphragm so that critical illumination can be achieved (Koehler illumination) .

The Selection and Use of a Compound Microscope

A recommended compound microscope suitable for studying fish chromosomes should consist of the following components:

Fluorite objectives (10X and 40X)
Planapochromatic oil immersion objective (N. A. 1.4)
Compensating eyepieces (10X)
Aplanatic-achromatic condenser (N. A. 1.3-1.4)
A built in 6 volt illuminator with variable transformer and field diaphragm
A trinocular stand (binocular eyepieces and a phototube) sturdy enough to support the weight of a camera assembly)
A graduated mechanical stage

Routine scanning for the locations of stained metaphase spreads is done at 100X or, if the chromosomes are unusually small, at 400X. Fluorite objectives with compensating eyepieces provide suitable clarity for this purpose. The oil immersion objective is used for the most critical examination of chromosomes. The planapochromatic objective provides the best clarity and the higher cost of this item is well justified when the end result is considered. A quality oil with a high refractive index should be used with this objective. A recommended procedure to obtain optimum clarity is given below. This will be time consuming in the beginning but after a few trials the necessary settings can be made very quickly. It is especially recommended before taking photomicrographs. This procedure is for a research microscope with a built in illuminating system using the oil immersion objective.

1. Align the light filament with the optical path of the objective by closing the condenser diaphragm, and opening the field (light source) diaphragm, and then focusing the image of the light filament on the closed leaves of the condenser diaphragm.
2. Place a slide containing stained chromosomes on the microscope stage and obtain an image under oil.
3. Open the condenser diaphragm and close the field diaphragm leaving a small circle of light.
4. Adjust the condenser up or down until the edges are in sharp focus, then open the field diaphragm until its edges just disappear from the field.
5. Remove the eyepiece and close the substage diaphragm until about three fourths of the field is open. If the condenser is off center, it may be aligned by adjusting the centering screws located on its sides.
6. Replace the eyepiece and vary the intensity of the light field. A green filter inserted between the condenser and light source will improve definition both visually and photographically. If the light is too bright, neutral density filters should be used since loss of definition will result if the above settings are varied.

The above adjustments to properly columnate the beam of light is termed Koehler illumination. This will provide the best lighting to distinguish true shape or size. It may not, however, provide optimum contrast. Contrast means the relative difference in intensity between the object and its immediate surroundings. The limit of detection is set by the intensity of the light and the contrast. In enumerating chromosomes this is all important, and resolution without contrast is of little use. Sometimes a little resolution may have to be sacrificed in order to obtain maximum differences in intensity, but if the chromosomes are well stained and/or if the proper filters are used, then Koehler illumination will usually provide both maximum resolution and contrast.

PHASE MICROSCOPY

One of the most valuable pieces of equipment for examining chromosomes is the phase contrast microscope. Essentially, this is a modified bright field microscope for viewing unstained, or faintly stained objects. A phase scope has all the attachments of the ordinary bright field scope, but in addition, it has a sub-stage phase annulus below the condenser, and a diffraction plate in the rear of the objective. These two additions accentuate phase changes in the light waves entering and emanating from the specimen and is perceived by the eye as shades of light and dark.

The light microscope can be converted to phase by changing the condenser and objectives. Some condensers are sold as both bright field and phase but at high magnifications there is some loss of clarity. Phase optics have three distinct advantages in studying chromosomes. It provides a means for scanning and evaluating slides that are unstained or poorly stained; it enhances the contrast of stained material; and, it gives a uniform intensity to all the chromosomes within a given complement which is especially desirable when making photomicrographs for preparing karyotypes.

THE INVERTED MICROSCOPE

The inverted microscope, often referred to as a tissue culture microscope, is optically designed to view objects inside thick containers. The objectives are located beneath the stage and the condenser is positioned from two to six inches above the specimen. It may be purchased with bright and/or phase optics. The effective magnification range is usually from about 30x to 400x. In fish cytology, the inverted microscope is used mostly for viewing the growth of fibroblasts in tissue cultures. The lower cost instrument with phase attachments is sufficient for this purpose although more expensive microscopes with special condensers and oil immersion objectives are available. Special containers made of optically clear glass must be used with the latter. If the purpose is to view living chromosomes at high magnification, it would be better to use a regular phase compound microscope in conjunction with a special culture vessel, such as the Sykes Moore growth chamber, than to use an inverted microscope.

PHOTOGRAPHY OF CHROMOSOMES

The Photomicroscope

It should be emphasized again that the research microscope is all important since it is the primary means for obtaining a crisp, clear image of a chromosome spread. Any camera assembly used with the microscope will merely record that image. The best system for doing this is one that is built into, or has been adapted to, the optical system of the microscope. Basically this consists of a camera back, a viewing tube, a way to divert the light path to and away from the camera, and a means for taking a light reading. The last item is not absolutely necessary since the amount of light for a correct exposure can be determined through trial and error. The market is saturated with all types of systems containing these items. The choice of which type of camera back to use is an important one. For routinely taking large numbers of pictures, it is faster and less expensive to use a 35 mm camera system. On the other hand, the size of the image recorded on the negative is small. Also, if a mistake is made in exposing the negative, it won't be detected until the film is developed and all the figures will have to be relocated. The other choice is a 4 x 5 inch camera back with a bellows extension. This system uses sheet film which will record a much larger image. It also has an additional advantage. The film can be developed and evaluated immediately after the picture is taken without having to move the stage setting. If a mistake is made, another photograph can be taken immediately. This system however, takes more time and sheet film is more costly. Many photomicroscopes have both systems. Either the 35 mm or the 4 x 5 inch camera back can be used interchangeably.

The less expensive camera systems on the market are manually operated. These can usually be attached to any microscope and are quite suitable for photographing chromosomes. The more expensive systems are automatic and may, or may not, be built into the microscope. These are convenient to use and provide more exact exposures in a shorter length of time. Whichever system is used for taking photographs through the microsocpe, Koehler illumination should be established to produce the best possible image at the level of the film.

Film

For 35 mm photography, Plus-X panchromatic (Eastman Kodak) is an all purpose film for chromosomes. This film has medium speed, fine grain, good contrast, and is moderately priced. Money can be saved by purchasing one hundred feet rolls and loading individual film cassettes with as many frames as needed for a particular series of exposures. Contrast Process Panchromatic, or Orthomatic (Eastman Kodak), is the most recommended 4 x 5 inch sheet film for photographing chromosomes. This film has medium speed, good resolving power and very high contrast. It is sold in lots of twenty-five or one hundred sheets per box.

Developing Film

After photomicrographic exposures have been made, the next step is to process the film. The standard procedure includes developing, rinsing, fixing, washing and drying. For 35 mm film these procedures are normally carried out in a small reel-type tank that allows liquids to be poured in and out while excluding light. The following procedure is carried out at room temperature.

1. In total darkness, remove the film from its cassette. Roll the film on a reel and place into a stainless steel developing tank. Most tanks will accept three separate reels. Place the cap on the tank. The lights may now be cut on for the remainder of the developing procedure.
2. Pour D-11 developer (Eastman Kodak) full strength or dilute with water (1:1) through the light-tight opening in the developing tank.
3. Develop for five minutes. The tank need not be agitated during this time.
4. Pour off the developer and rinse with tap water.
5. Fix in Kodak Fixer (Eastman Kodak) for five minutes.

6. Pour off the fixer and wash under running water for at least twenty minutes.
7. Pour off the water and rinse in Photo-Flo (Eastman Kodak) diluted with water (1:200) for thirty seconds. This step helps prevent spotting but is optional.
8. Pour off the Photo-Flo, unwind the film from the reel and hang up to dry.

For 4 x 5 inch sheet film, the following procedure is recommended at room temperature:

1. In total darkness, remove both sheets of film from the 4 x 5 inch film holder and place in an open tray containing D-11 (full strength or diluted 1:1).
2. Develop for five minutes without agitation.
3. Remove the film from the tray, rinse under running water and fix in Kodak fixer for five minutes. The lights may now be turned on.
4. Take the film out of the fixer and wash in running water (preferably a swirl bath) for at least twenty minutes.
5. Place in Photo-Flo for thirty seconds and hang up to dry.

Printing

Black and white protomicrographs can be printed from negatives to positives by projecting the image on the negative to positive paper, then developing and fixing the positive. The following is one recommended procedure to be carried out at room temperature.

1. Insert the 35 mm or 4 x 5 inch negative into a photographic enlarger.
2. Focus the image to be developed on an easel containing an exposed sheet of Kodabromide F_2 print paper (Eastman Kodak). This can be done in total darkness or under a safelight.
3. Once a sharp image is obtained with suitable contrast, turn off the enlarger light and replace the print paper with an unexposed sheet (emulsion or glossy side up).
4. Set the automatic timer and expose the print paper for a determined period of time.
5. Remove the paper from the easel and develop in a tray containing Dektol (Eastman Kodak) diluted with water (1:2). This is done under a safelight until a image of suitable contrast becomes visible.
6. Rinse the print under running water and place in fixer for five minutes.
7. Wash the print in a swirl bath for at least one hour and place on a

heated dryer with the emulsion side down. This will produce a glossy print.

Microscope Filters

Filters may be used with the microscope to improve the contrast of chromosomes when examining with the eye, or when taking photomicrographs. Listed below are some colors of filters that are recommended for improving resolution when certain colored stains are used. Those that are recommended for use with Giemsa and aceto-orcein stain are emphasized. More detailed information can be obtained elsewhere (Kodak publication No. B-3).

Stain Color	*Filter Color*
Violet	Yellow
Purple	Green
Blue (Giemsa)	Red (Wratten No. 25* or No. 29*)
Green	Red
Yellow	Blue
Red (Aceto-orcein)	Green (Wratten No. 58*)
Brown	Blue

*Eastman Kodak

The Camera Lucida

The camera lucida is a device that is used in conjunction with the microscope to superimpose the field image on a drawing surface. The older types fit onto the eyepiece and some practice is needed to get a distortion free image using oil immersion. The instrument consists of a partly silvered prism placed over the eyepiece of the microscope and a mirror for viewing the drawing surface. Some of the light passes from the microscope through the prism to the eye, and light from the drawing surface is reflected by the mirror through the prism to the eye. The result is an image of the pencil point superimposed on the image of the object. Contrasts are varied by selecting filters located around the prism, and by regulating the light around the drawing surface.

The newer types are built into, or can be fitted into, the microscope between the objectives and eyepiece This provides a brighter and more distortion–free image and can be left in place when not in use. The camera lucida has value in that microscope images can be correctly drawn to scale. It is also useful in depicting chromosomes of a complement that occur in more than one focal plane.

Measuring Chromosomes Through the Microscope

Measuring chromosomes through the microscope is accomplished with the aid of an ocular scale or reticule and a stage micrometer. Variously ruled scales may be used in any eyepiece. The top lens of the eyepiece is unscrewed and the reticule is placed with the ruled side down so that it rests on the diaphragm within the eyepiece. It is more expedient to have a separate eyepiece with a reticule already in place. When a chromosome is to be measured, the regular ocular is replaced with that containing the reticule. This also eliminates the possibility of damaging or scratching the expensive lens of the compensating eyepiece. The scales of the reticule have arbitrary units. Consequently, it must be calibrated for use with a stage micrometer which has definite units of measurement. The smallest of these units is usually 0.01 mm. Chromosome measurements are usually made only with oil immersion.

1. Place the stage micrometer on the microscope and focus with the oil immersion objective.
2. Replace the compensation eyepiece with the ocular containing the graduated reticule.
3. Rotate the eyepiece until its scale is either superimposed on, or very close and parallel to, the stage scale so that the zero mark of both scales coincide.
4. Record the number of divisions on the stage micrometer between the zero mark and a point on the opposite end of the scale where a line from each scale is superimposed.
5. Multiply this number of divisions by the true length of each space. Convert to microns and divide by the equivalent number of spaces on the eyepiece scale. This value represents the number of microns in one space of the ocular scale. Example: If 100 spaces on the ocular scale = 65 spaces on the stage scale with each space equaling 0.01 mm, then 100 spaces on the ocular scale = 0.65 mm or 650 microns. Therefore, one unit on the ocular scale = $\frac{650}{100}$ or 6.5 micons.
6. The ocular scale is now calibrated in definite micron units. Chrom-

osomes can be measured directly by determining the number of ocular units it takes to cover an individual chromosome dimension.

A screw, or Filar (Bausch and Lomb) micrometer may be purchased which consists of a measuring line that is moved by a control at the side of the eyepiece. This type of micrometer simplifies measurements because of a direct reading on an outside scale rather than counting the number of divisions as with the ocular micrometer. It has to be calibrated, however, in the same way described above for the eyepiece micrometer.

METHODS AND DEVELOPMENTS NEEDED IN STUDYING FISH CHROMOSOMES

Mammalian chromosome methodology consists of methods that have been developed over the past ten years that are currently not in use in studying fish chromosomes. These methods have been used with mammals to better identify individual chromosomes and to provide information about the cell cycle in general. A few of these methods are presented here to emphasize their importance and potential application to fish systems.

Giemsa Banding of Chromosomes

The differential staining properties of Giemsa stain were first noted by Pardue and Gall (1970) in studying mouse chromosomes. Since then, several methods have been reported for obtaining banding patterns with human chromosomes and other mammalian organisms. The qualitative effect produced with Giemsa is an alternation of light and dark bands, similar to the staining pattern obtained with Drosophila salivary gland chromosomes. The exact molecular mechanism for the banding is not known but it is thought that denaturation and renaturation of DNA is involved. Sumner, *et al.* (1971) have proposed that repetitive DNA along the chromosome is not denatured as easily as nonrepetitive DNA and becomes more solidly tempered to the slide during air drying. The Giemsa stain then binds preferentially to the undenatured DNA. A few of the techniques for banding chromosomes with Giemsa are given below.

ALKALINE GIEMSA FOR BANDING HUMAN CHROMOSOMES

(Patil *et al.*, 1971)

1. Stain air-dried chromosome preparations for 5 minutes with Giemsa (2 ml of Haleco Giemsa Blood stain azure A stock solution, 2 ml of 0.14 M Na_2HPO_4 and 96 ml H_2O) at a pH of 9.0.
2. Wash in running tap water for one minute and dry in air.
3. Observe without a coverglass or clear with xylene and mount.

Comments: Cells that are in mid-metaphase provide the best staining patterns for matching homologous pairs. Cells in early or late (contracted) metaphase are unsuitabe for homolog identification.

GIEMSA BANDING OF HUMAN CHROMOSOMES TREATED WITH TRYPSIN

(Wang and Fedoroff, 1972)

1. Treat air dried chromosome preparations with trypsin solution (0.025-0.05% trypsin in Ca and Mg free balanced salt solution) or with trypsin-versene (1 part 0.025-0.05% trypsin and 1 part 0.02% EDTA at pH 7) for ten to fifteen minutes at 25-30°C.
2. Rinse in two changes each of 70 percent and 100 percent ethanol and air dry.
3. Stain for one to two minutes in the following Giemsa stain: 5 ml stock Giemsa solution (Fisher Scientific Co.), 50 ml distilled water and 1.5 ml of 0.1 M citric acid solution adjusted to pH 7 with 0.2 M Na_2HPO_4.
4. Rinse twice in distilled water, air dry, and mount.

Comments: It is assumed that the trypsin removes the protein component of the nucleoprotein complex, which has been denatured by fixation, and allows the Giemsa stain to react directly with the nucleic acid moiety. If the trypsin treatment is too short or too long, clear banding patterns are not observed.

GIEMSA BANDING PATTERNS OF CHINESE HAMSTER CHROMOSOMES

(Kato and Yosida, 1972)

1. Within twenty-four hours after air drying chromosome preparations, incubate slides in 2 M NaCl-5M urea (pH 6-7) for twenty minutes at 37°C.

2. Rinse briefly with tap water and stain with Giemsa diluted about forty times with Sorensens' buffer (pH 7.0) for five minutes.
3. Rinse in tap water and air dry.

Comments: Banding was unsuitable when preparations were flamed. The best results were obtained when the slides were dried overnight at room temperature.

Fluorescent Microscopy

Fluorescent microscopy involves the use of special dyes (fluorochromes) and ultraviolet light to produce a banding effect on chromosomes similar to that produced with Giemsa stain. When the fluorochrome-chromosome complex is radiated with light of one wavelength, it will reradiate light of a longer wavelength and fluorescence is observed. The fluorochrome binds preferentially to the nucleic acid and results in light and dark bands. This effect has been produced mostly with metaphase chromosomes from human leucocytes. In addition, valuable information has been obtained with fluorescence of interphase cells. The human Y chromosome, for example, will fluoresce much brighter than other chromosomes in the complement providing a means to identify sex without resorting to a complete karyotypic analysis.

An added dimension in evaluating the fluorescent bands is the enumeration of distinctive photoelectric absorbance curves by scanning banned chromosomes. These curves measure the relative absorbances of the ultraviolet rays and provide additional data to better match up homologs and identify structural aberrations.

A number of fluorochromes are available for use with fluorescent microscopy. Quinacrine mustard and the antimalarial drug quinacrine dihydrochloride (Atabrine) are the ones most widely used to date. Caspersson, *et al.* (1970) were the first to emphasize the value of fluorescent banding with plants and later with humans. His method is briefly presented as follows:

FLUORESCENT BANDING OF HUMAN CHROMOSOMES
(Casperson *et al.*, 1970)

1. Air dried preparations of leucocyte chromosomes are placed in absolute ethanol for five minutes then in 96 percent, 75 percent, and 65 percent ethanol for approximately one minute each.

2. Place the rehydrated preparation in MacIlvaine's disodium phosphate/citric acid buffer at pH 7.0 for five minutes.
3. Stain in a buffered solution of quinacrine mustard (50 mcg/ml of quinacrine mustard in MacIlvaines buffer) for twenty minutes.
4. After staining, rinse with distilled water. A temporary mount is made by placing a drop of distilled water on the preparation, applying a coverslip, and ringing with paraffin (permount and other mounting media is unsuitable because of their fluorescent properties). This should be good for two to three days.
5. Observe under a fluorescent microscope.
6. Photographs are made with Tri-X panchromatic 35 mm film (Eastman Kodak).

Comments: An alternative flurochrome stain is quinacrine dihydrochloride (Atabrine). This is used as a 0.5 percent aqueous solution at a pH of 4.5. This stain is much cheaper than quinacrine mustard and is easier to prepare. A number of fluorescent microscopes are available although the ones manufactured by Leitz and Zeiss are most commonly used with chromosomes. The Zeiss assembly carries the following components: An HBO 200 W mercury lamp; BE (2.5 mm) and OG 1 (1.5 mm) filters; an Epicondenser with a dichroitie reflector (limiting wavelength approximately 480 mm); and a short wave cut-off filter (limiting wavelength of approximately 500 nm) for suppression in the fluorescence beam.

Several attempts are usually necessary before acceptable bands are obtained. Giemsa banding is far less expensive and can be done in a shorter period of time. Also, fluorescent banding preparations tend to fade under ultraviolet light.

Electron Microscopy

One of the most significant findings of recent times has been the elucidation of the ultrastructure of metaphase chromosomes. Compared with the advances in chromosome chemistry and the ultrastructure of cytoplasmic organelles, the determination of the fine structure of metaphase chromosomes has lagged far behind.

This has been due primarily to the technique for preparing chromosomes for transmission electron microscopy. Electron microscopy differs from light microscopy in that electrons, instead of light, are beamed from an energy source, through a specimen, to an observation screen or photographic plate. The usual method for preparing specimens is to embed fixed tissues in a plastic or resin medium, preparing thin sections with an ultramicrotome, then positioning the sections on a small grid for observation. The electron beam must operate in a vacuum which means the specimen must be dry. Exceptionally thin sections must be used and sometimes atoms of metal are used to increase contrast. This method has provided valuable data on the ultrastructure of prokaryotic nuclei and the spindle apparatus of eukaryotic cells, but because of the diffuse and porous properties of metaphase chromosomes, insufficient contrast is produced to enumerate their fine structure.

In 1965, Dupraw (Dupraw, 1965) reported a special technique for viewing the ultrastructure of chromosomes from the embryonic cells of the honey bee. This showed a substructure of individual fibers two hundred to three hundred Angstroms wide arranged in a folded network. In 1966, the same investigator (Dupraw, 1966) applied the same technique to human metaphase chromocomes (Fig. 3-1). In addition to the small coil of fibers found in the honey bee, the chromosomes exhibited macrocoiling and interconnecting fibers. A number of studies have since been done to substantiate and better clarify the image of these structures.

DuPraw's technique consists of preparing chromosomes as whole mounts rather than as thin sections. Chromosomes from growing leucocytes or some other cultured cells are blocked at metaphase with colchicine and spread on air-water interfaces. This causes the cells to burst and release the chromosomes. The intact chromosomes are then picked up on electron microscope grids, washed or treated with reagents, then dried from liquid carbon dioxide by the critical point method described by Anderson (1951). The application of this last step is a very important one and it has been the single most factor in the success of prepar-

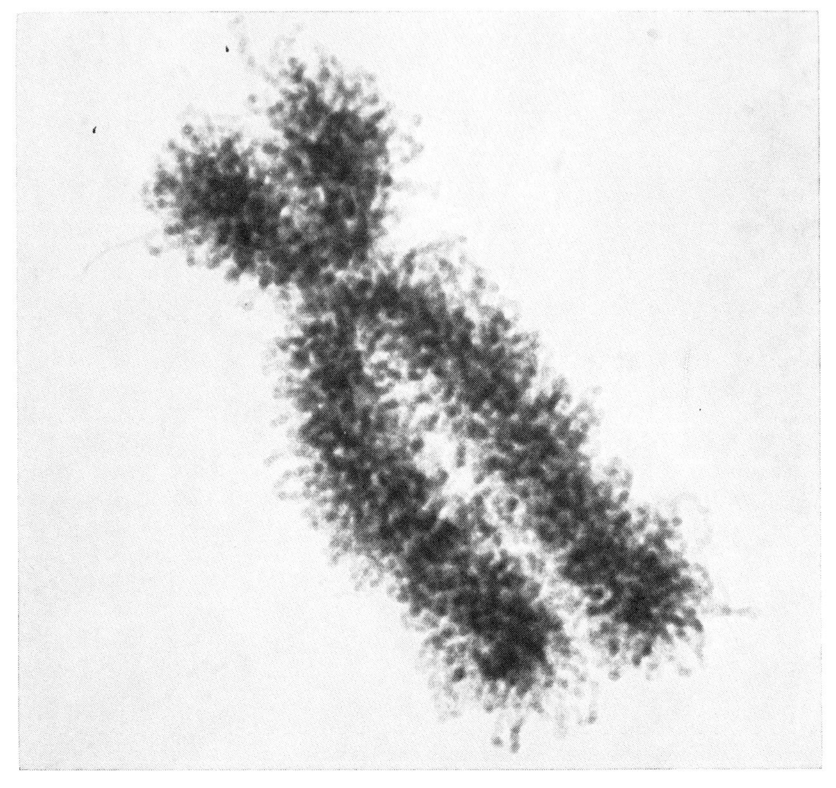

Figure 3-1. The ultrastructure of a typical human metaphase chromosome showing loops or folded fibers. DuPraw (1970) *"DNA and Chromosomes,"* Holt, Rinehart & Winston, N.Y.

ing intact whole-mount chromosomes for electron microscopy. This method preserves the characteristic structure seen with the light microscope but the magnification and resolution are much greater.

Much could be learned about fish chromosomes using this method for studying human chromosomes. Some of the possibilities for exploration would include: The arangement of DNA within the fiber by comparing the total length of the DNA double helix with the total length of the fiber (DNA packing ratio) ; the nature of the arrangement of fibers in specific chromatids; the elucidation of centromeric and satellite regions; the enumerations of interconnecting chromosome fibers (multivalent association); the position of chromosomes on the metaphase plate; the identification of sex chromosomes; the substructure and relationships of nucleoli; the synaptic arrangements and tetrad associations in meiotic cells; and the ultrastructure of chromosome aberrations with and without treatment with specific chemicals.

The diversity of chromosome types found in fishes would provide excellent source material for studies such as those given above. There are the microchromosomes of the lamprey, the extra large chromosomes of the lung fish, the presumed sex chromosomes of *Gambusia,* and readily available testicular material in a number of forms. Studies such as these are very much needed and would be a welcome contribution to the field of fish cytology.

Autoradiography

Autoradiography is a photographic study of the effects of a radioactive isotope in a biological system. In studying chromosomes, the isotope is normally incorporated into the DNA which is then localized with a photographic emulsion. Ionizing radiations, like light photons, produce images on film emulsions that can be transformed into visible images by normal photographic developing and fixing procedures.

The value of using this system to study chromosomes was first revealed by Taylor, *et al.* (1957) when tritiated thymidine was administered to the chromosomes of *Vicia faba.* Later on a number of investigators demonstrated asynchronous DNA replication patterns in animal cells. Since then much information has accumulated on specific times of division for individual chromosomes, time intervals for the cell cycle, and the identification of specific chromosomes.

In practice, actively dividing cells are exposed to an isotope, commonly tritiated thymidine, for a definite period of time. The isotope may be replaced with unlabelled medium after a few hours (pulse labelling), or left for a longer time until the cells are to be harvested (continuous labelling). Either way, the cells are exposed to colchicine, collected, treated with a hypotonic solution, fixed and air dried. The slide preparation is then coated with a photographic emulsion. Because of its small size, commercial film cannot be used. Stripping film, a preparation in which the protective backing can be removed from the emulsion; or liquid emulsions, which dry rapidly after dipping to form a coat of sensitive material, are most commonly used. The application of film is done in the dark. An appropriate time is allowed for the beta radiation from the incorporated thymidine to expose the silver in the photographic emulsion. Development with photographic developer discloses dark grains which are superimposed over regions of the chromosome. When used with conventional staining procedures, the location and intensity of the isotope can be determined. The working procedure for chromosome autoradiography will not be given here. The reader is referred to Priest (1969) for a thorough account of the technique.

The cell cycle of eukaryotic organism consists of a complex interphase (G_1, S, and G_2) and an M stage which includes prophase, metaphase, anaphase, and telophase. The introduction of radioactive thymidine at these various stages would provide useful information about the length of the cell cycle, and the degree of asynchronous replication of DNA in various chromosomes. Some of the chromosomes of a complement are labelled differently from all the others. The X chromosome of humans, for example, is the last to replicate and labelling with thymidine is the only true way to identify this chromosome. The chromosome complement of fishes appear to be very heterogenous. No doubt autoradiography would elucidate some interesting and worthwhile findings in complements such as these since information about time relationships of DNA replication and separation is unknown for this group. Also, because of the availability and habitat of fishes, *in vivo* autoradiographic studies could easily be designed. Many compounds can be purchased in isotope form. These could be administered to the fish and the effects on chromosomes resulting from their incorporation could be studied using autoradiographic methods. Regardless of whether the studies are done *in vitro* or *in vivo,* the use of isotopes in studying fish chromosomes should prove to be a rewarding endeavor.

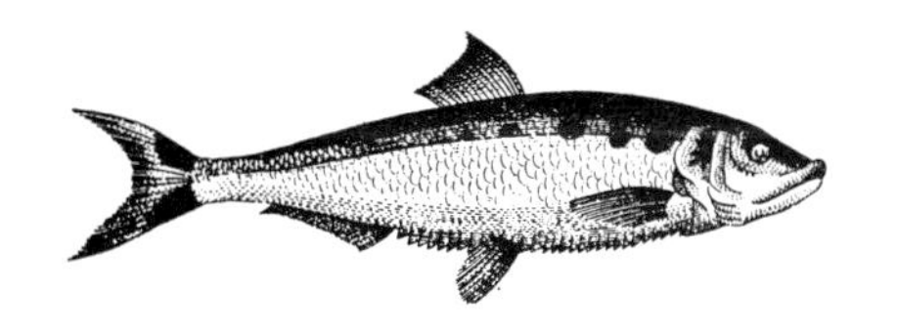

CHAPTER 4

THE FISH KARYOTYPE

BY THE TIME MITOSIS has proceeded to metaphase the chromosomes have reached their most distinct and characteristic form. It is in somatic metaphase that chromosomes are best evaluated or karyotyped. A karyotype is the characterization and analysis of a chromosome set within the nucleus of a given species. This includes defining the chromosome size, type, and all morphological characteristics. The fish karyotype has been harder to define than that of any other vertebrate. The human karyotype was the first to be studied in depth and a committee, meeting in Denver, Colorado in 1960, established a format and style of presenting human metaphase chromosomes that is now universally followed. Unfortunately, many fish cytogeneticists have tried to adopt this system for the presentation of fish chromosomes without taking into account that humans are represented by only one type of complement whereas the complements of fishes vary with each species studied. Various workers present fish karyotypes in their own way which is reminiscent of early works with human karyotypes. Consequently, no convention is followed. This presents difficulties when the investigator tries to relate his works with the works of others. The fish karyotype and its many ramifications is the subject of this chapter.

Chromosome Structure and Number

The chromosomes of fish complements include autosomes and sometimes sex chromosomes or heteromorphic pairs that may or may not be sex chromosomes. Autosomes constitute the most common or typical chromosomes and make up the majority of the complement. Sex chromosomes are much less numerous and because they are somewhat undefinable, they will be treated separately later on in the chapter. Autosomes are composed of two strands or chromatids joined together by a centromere. It is generally concluded that the chromatid is nucleoprotein in nature and can coil, producing contraction, or uncoil, resulting in elongation. Light microscopists have described chromosomes for decades and have coined terms to describe what they saw when various stains or chemicals were applied to cells of both plants and animals. These studies, however, have not resulted in an accurate description of the chromosome's true structure. It was mentioned earlier that electron microscopists have recently made headway in defining the submicroscopic parts of mammalian chromosomes. No one has reported on the ultrastructure of fish chromosomes, but there is no reason to doubt that their basic autosomal structure is any different from that of mammals. In essence, the chromatid is composed of folded fibers two hundred to three hundred angstroms wide which are arranged in metaphase to produce the image that we recognize as a two stranded structure with the light microscope. Some of these fibers may have broken ends, which could account for the *stickiness* often associated with some chromosomes, and they may be interconnected with fibers of other chromosomes. Within the fiber is the DNA-protein having a diameter of about one hundred angstroms. The coiling of these fibers are dependent on the presence of histone protein and divalent cations. The possibility that such a chromosomal model exists for all organisms is highly intriguing. If this is indeed the physical structure in which DNA resides for all eukaryotic cells, then the study of DNA transferance in higher organisms will be boosted considerably in the years to come as their fibers become more clearly defined.

For more than a century it has been recognized that cells contain more than one chromosome. With increasing certainty it has been shown that species are characterized by a definite chromosome number or set. The haploid ($1n$) set contains the gametic number of chromosomes, or those chromosomes transmitted to offspring. This set of chromosomes is also termed the genome. The diploid set ($2n$) constitutes the zygotic number or two parental haploid sets. Each autosome, and usually sex chromosome, of the diploid complement has a partner or homologue. In fishes, both haploid and diploid sets contain chromosome numbers characteristic for the species. The chromosomes of a set may or may not vary with respect to size and shape. Furthermore, chromosome sets may vary between two closely related species, or there may be close similarity between chromosomes of distantly related forms. Chromosome sets in fishes range in number from a $2n$ of 16 in some tropical species, to a $2n$ of 174 in some lampreys. The sizes of fish chromosomes are also highly varied. The microchromosomes of the lamprey are dot like and less than 1 micron in length whereas some chromosomes of the lungfish measure about 30 microns.

Meiosis in fishes is similar to that of other vertebrates. There are two divisions in which there is only one chromosome duplication and the nucleus and cytoplasm divides twice to produce four daughter cells. These two divisions are termed meiosis I and meiosis II. At the end of meiosis I the homologous or parental chromosomes have separated so that each of the two cells that are produced contain one-half the original number. At the end of meiosis II are four daughter cells that contain single stranded chromosomes of the reduced number. Meiotic chromosomes are most clearly seen during metaphase's I and II. Before metaphase I the homologous chromosomes have paired or synapsed during prophase I, forming four stranded tetrads or bivalents. These bivalents become more discernible as the chromosomes contract and separate toward their alignment in metaphase I. In some cases more than two homologous chromosomes will synapse forming trivalents, quadrivalents, etc. Synapsed chromosomes other than bivalents may be referred to collectively as multivalents, particularly if the exact number of grouped chromosomes cannot be ascertained. Characteristically, these appear as long rods or large O-shaped structures. The chromosomes seen during metaphase II are rod shaped and single stranded. These are easier to count but the similarities and differences of chromosomes cannot be evaluated as well as those in metaphase I. Very few workers karyotype meiotic chromosomes. With existing means it is difficult to identify all specific types and to coordinate them with those of somatic spreads. For this reason a photograph of metaphase I chromosomes is usually all that is needed to point out obvious multivalents or unusual synaptic configurations. If the meiotic chromosomes are large and few in number then a more detailed means of presentation would certainly be in order.

Chromosome Nomenclature

The position of the centromere is the basis for the classification of a chromosome. The centromere is a structural and functional subunit that provides a site for the attachment of spindle fibers during chromosomal movements. It is also referred to as the kinetochore or primary constriction. If the centromere is approximately in the center of the chromosome, its position is said to be *median*. If the centromere is at the end of the chromosome, its position is *terminal*. If it is located approximately midway between the center and the end of the chromosome it is *submedian*, and if located between the terminal and submedian position, it is *subterminal*.

The chromosome type is the morphological characterization of a chromosome based on the relative position of its centromere. If the centromere is median, the chromosome type is said to be *metacentric*. If the centromere is submedian, its type is *submetacentric*. If the centromere is terminal the type is *acrocentric* or *telocentric* and if the centromere is subterminal, the chromosome type is *subtelocentric*. Sometimes chromosomes with terminal and subterminal centromeres are both described as acrocentric.

Although it is seldom used anymore, an older terminology for describing chromosome types is that based on anaphase configurations. These are denoted as V-shaped, J-shaped, or rod-shaped chromosomes and correspond to metacentrics, submetacentrics, and acrocentrics respectively. All three of these relationships describing centromeric position and chromosome type are summarized as follows:

Centromeric Position	*Chromosome Type*	*Anaphase Configuration*
Median	Metacentric	V-shaped
Sub-median	Sub-metacentric	J-shaped
Terminal	Acrocentric (or Telocentric)	Rod-shaped
Sub-terminal	Sub-telocentric	Rod-shaped

The most evident and accurate way to depict a chromosome is to relate whether or not its centromere is median, submedian, subterminal, or terminal. For some unknown reason it is improper to refer to a chromosome in terms of its centromeric position. Levan, *et al.* (1964) suggested that the chromosomes with these centromeric types be designated as *m, sm, st,* and *t* chromosomes respectively. In this way anyone would have full knowledge of the description of any chromosome. This proposed system was not adopted by the masses. Instead, the terms metacentric, submetacentric, and acrocentric have become well indentured into the literature principally as the result of the characterization of mammalian chromosomes. As a consequence, one is compelled to work with the system as it has evolved and since most fish cytogeneticists prefer to use these terms, they must be reconciled.

Even when centromeric positions and chromosome types are related, there are still points of confusion in visually grouping chromosomes due to regions of overlap. This confusion may be eliminated by assigning definite numerical values for each category of chromosome type based on arm ratios. Levan, *et al.* (1964) proposed such a system of values based on the ratio of the length of the long arm divided by the length of the short arm. This numerical designation is workable, reproducible, and very much recommended for classifying fish chromosomes. Table *I* gives these values relative to the centromeric positions and chromosome types already mentioned. It is further proposed that the terms metacentric, submetacentric, subtelocentric, and acrocentric be used to denote chromosome type and that the long arm-short arm ratios (L/S) proposed by Levan be used as points of separation. The only real problem in past years has been the designation of chromosomes with centromeres located close to the end. If no short arm is detectable, or if one exists but is too short to measure, it should be termed acrocentric. If there is a measurable short arm, falling in the L/S range of 3.01-7.00, it should be classified as subtelocentric.

The Photokaryotype

A photokaryotype is the systematic arrangement of the chromosomes from a photograph representing one nuclear spread. The

TABLE I
RECOMMENDED NOMENCLATURE FOR DESIGNATING CHROMOSOME TYPE*

Centromeric Position	*Arm Ratio* (L/S)	*Chromosome Type*
Median	1.00-170	Metacentric
Submedian	1.71-3.00	Submetacentric
Subterminal	3.01-7.00	Subtelocentric
Terminal	7.01-	Acrocentric

*Taken from Levan, *et al.* (1964).

metaphase chromosomes from a photographic print, which is representative of a full diploid complement of the species, are cut out and arranged to show comparable sizes and shapes. If the chromosomes are large and distinct, this can usually be done visually. The more exacting method is to measure the overall length of the chromosomes and the length of both arms before grouping. If the chromosome has only one arm then only a single measurement is necessary. In preparing a fish photokaryotype, the following method is recommended:

1. Cut out the individual chromosomes from a photographic print with good contrast. The print should be as large as possible without loss of definition. Scissors may be used but it is faster and easier to staple the print to a stiff piece of cardboard and cut out the chromosomes with a sharp pointed scapel.
2. The chromosome cut outs should be placed in a container, such as a petri dish, as soon as they are removed to prevent their loss.
3. Visually arrange the homologous pairs of chromosomes on a piece of white poster board of suitable size. They should be grouped ascording to like chromosome type in order of decreasing length e.g., all the metacentrics should be paired and grouped together from the largest to the smallest size. Sex chromosomes or odd chromosomes should be grouped separately at the very end.
4. For more accurate arrangements, each chromosome should be measured with dial calipers, or a similar instrument, and the values recorded on the back of each cutout. The measurements should include the overall length, the length of each arm, and the ratio of the length of the long arm divided by the length of the short arm (L/S). For acrocentric chromosomes, this last value would approach infinity. The resultant pairing of chromosomes can then be done numerically and objectively as well as visually.

5. Once the chromosomes are grouped in the final form they can be taped in rows with magic mending tape (Scotch brand). In order to prevent movement of the chromosomes due to static conditions the tape is slightly moistened with water. All edges should be covered and air bubbles removed with thumb pressure.
6. A line scale in microns should be placed along with the chromosome grouping to denote the relative lengths of each chromosome. This is done by measuring the longest chromosome through the microscope (Chapter 4), then measuring the same chromosome in the photograph and placing a line among the chromosomes that coordinates the two measurements.
7. The final karyotype may be photographed and printed to any size desired. The magnification of the final print is determined by dividing the length of the longest printed chromosome in microns by the true length of the chromosome as measured through the microscope.
8. An example of a photokaryotype prepared in this way is presented in Figure 4-1.

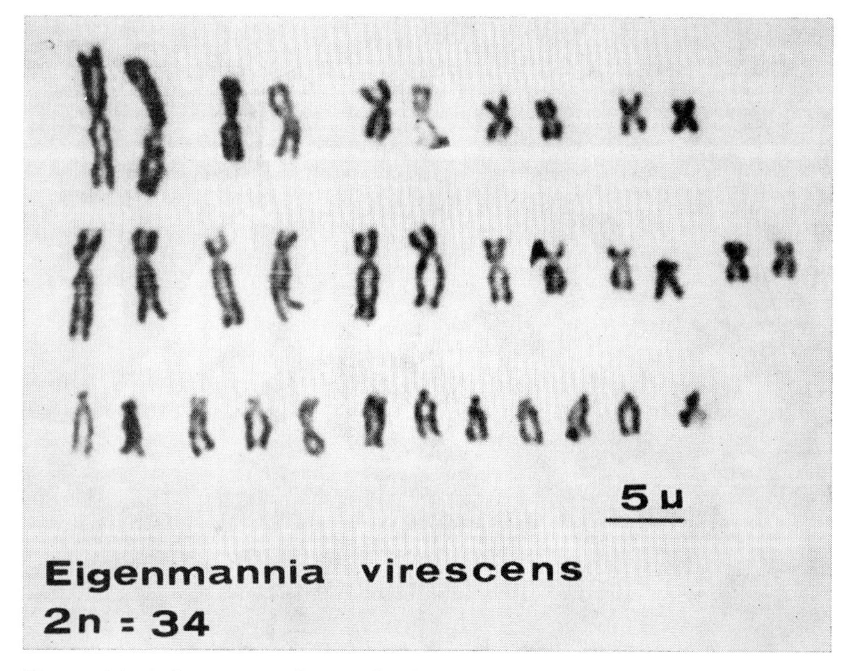

Figure 4-1. A karyotype of somatic chromosomes from the fin epithelium of *Eigenmannia virescens*. $2n = 34$.

The Eye Karyotype

The value of an eye karyotype is often underestimated and is seldom done. This is a free hand *birdfoot* drawing of a complement of chromosomes while viewing them directly through the microscope. The various types of chromosomes are scored and grouped together (Fig. 4-2). This enables the observer to use depth perception in defining chromosomes that are not well spread.

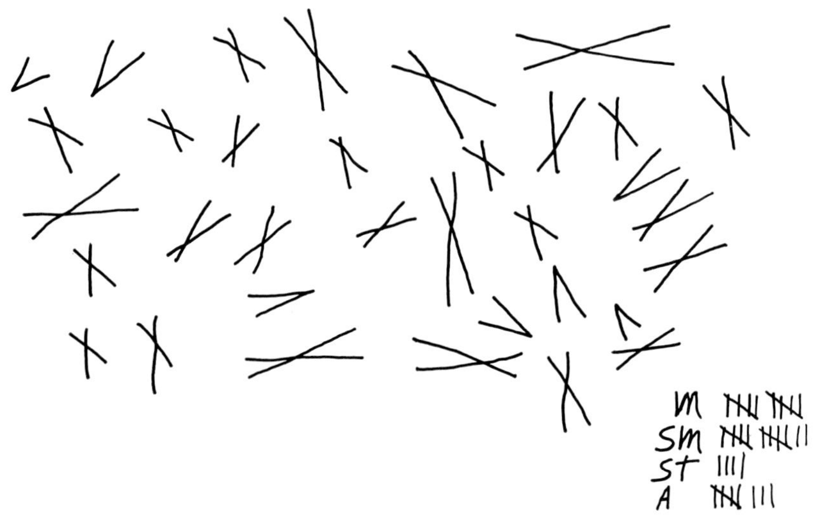

Figure 4-2. An eye karyotype of Figure 2-2. *(Eigenmannia virescens)*. This particular spread shows a $2n$ number of thirty-four and contains ten metacentrics (M), twelve submetacentrics (SM), four subtelocentrics (ST), and eight acrocentrics (A).

This also quickly familiarizes the viewer with specific chromosomes particularly when evaluating new complements. Eye karyotypes are time consuming at first but speed and proficiency are acquired through practice. Valuable corollaries are obtained when used in conjunction with photokaryotypes. In addition, the eye karyotype enables the investigator to score metaphases without resorting to photographing each and every spread.

The Idiogram

An idiogram is the arrangement of a haploid complement of chromosomes according to centromeric position and in order of decreasing length. This may be done from a photograph but is usually presented diagramatically in which the chromosomes are represented as straight lines. The idiogram has value when comparing representative complements of two or more species of organisms. To prepare an idiogram, it is necessary to match up homologous chromosomes and to subsequently prepare a diagram consisting of representative measurements for each homologous pair. An example of a fish idiogram is given in Figure 4-3.

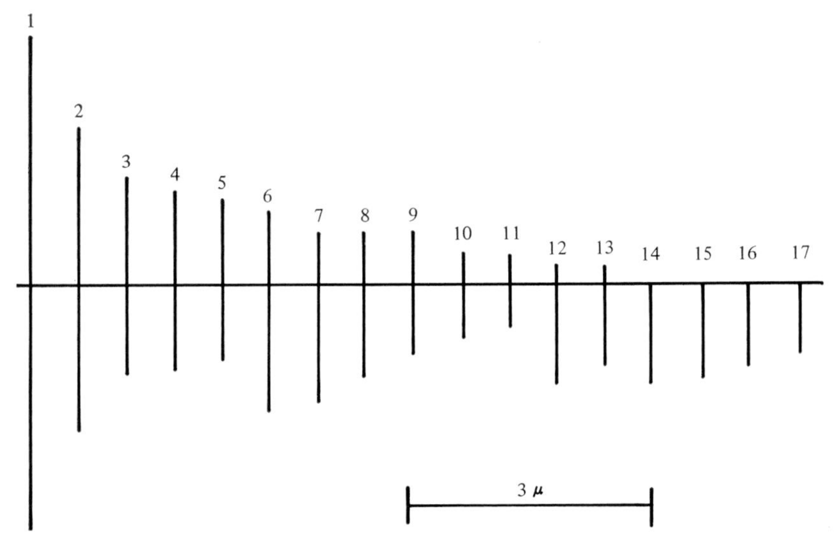

Figure 4-3. An idiogram of *Eigenmannia virescens* (inclusive of Fig. 2-2.) showing centromeric positions of the haploid complement. The above idiogram is representative of karyotypes from several different cells. Chromosomes 1-5 are metacentric, 6-11 submetacentric, 12-13 subtelocentric, and 14-17 acrocentric. With an idiogram, chromosome types, sizes and arm numbers can be easily compared with those representing other species.

Chromosome Modifications

Most of the time the chromosomes obtained from fishes will appear as two strands that are evenly stained with a centromeric region that stains lighter. Occasionally, some of the chromosomes of the complement will appear differently from the others. If these differences are consistently obtained then they are referred to as marker chromosomes and are valuable in characterizing specific karyotypes as well as providing a means for association with other karyotypes. Such chromosome modifications are important and should be noted with each metaphase karyotypic description.

A *secondary constriction* is a constriction in a chromosome in addition to the primary constriction of the centromere. It appears as a light staining region with a piece of chromosome extending beyond it. These extensions are termed satellites and the chromosome is referred to as an SAT chromosome. The satellites may vary in size but are usually very small and lightly stained. They are usually associated with acrocentric chromosomes. Some satellites may be associated with the nucleolus during interphase and have a function in its formation.

Megachromosomes are unusually large chromosomes in the midst of chromosomes of normal size. There may be only one per complement and seldom more than two. It is not known how they are formed. Oftentimes *macrochromosomes* are confused with megachromosomes. Unlike megachromosomes, macrochromosomes refer to the large chromosomes of a karyotype that possess only two sizes. One size consists of small dot-like chromosomes termed *microchromosomes.* The other size, the macrochromosomes, are larger and have a distinctive chromosome type. Within the macrochromosome complement the sizes may vary. A karyotype containing both machochromosomes and microchromosomes is said to be asymmetrical. In metaphase spreads the larger chromosomes tend to be located around the periphery.

Sometimes chromosomes stain differentially and are said to be *pyknotic.* That part of the chromosome that stains heavily is positively pyknotic and that which stains lightly is negatively pyknotic. Before characterizing a chromosome as pyknotic the differential staining pattern should be consistently reproducible. The reason for the different staining patterns is not known.

Supernumerary chromosomes are small chromosomes or fragments that appear in some complements in addition to the regular diploid set. It is thought that such chromosomes are genetically inert and unnecessary to the organism. Their function and mechanism of formation are not known. They are also referred to as accessory or B chromosomes.

In some fish karyotypes there are chromosomes of unlike size or shape that may be found to synapse together during meiosis. These may or may not be sex chromosomes. Some autosomes contain deletions, inversions or translocations that are not harmful to the organism. The altered chromosome pair is usually confined to cells of individual specimens. Some, however, could be characteristic for the entire species of a given population.

SEX CHROMOSOMES

Sexuality in Fishes

Before discussing sex determination and cytological evidence for sex chromosomes it is necessary to first define the various sex mechanisms found in fishes. According to Yamamoto (1969) sexuality in fishes covers an almost complete range of those types found in vertebrates. Essentially, this consists of various forms of hermaphrodism and gonochorism. If the organism possesses both ovarian and testicular tissues, the species is hermaphroditic. In turn, hermaphrodism may be either synchronous or consecutive. In synchronous (or balanced) hermaphrodism the male and female sex cells ripen at the same time. The testicular interstitial cells and ovarian follicular cells share a common embryonic blastema. The androgens and estrogens are some way kept separate and the mature fish completes both spermatogenesis and oogenesis at the same time. Synchronous hermaphrodism has been reported for members of the family Serranidae (e.g., *Serranus scriba*) and Cyprinodontidae (e.g., *Rivulus maroratus*). In *Rivulus marmoratus*, Harrington (1961) demonstrated that this hermaphroditic species was capable of internal self-fertilization.

In consecutive (or metagonous) hermaphrodism, species may be either protogynous, which may function first as females and then transform into males, or protandrous which transform from males into females. Protogynous hermaphrodism is found in the Sparidae (e.g., *Pagellus erythrinus*), and Serranidae (e.g., the sea bass *Centropristes striatus*). Ovarian lamellae make up most of the tissue in the gonadal cavity. Subsequently, seminiferous tissue begins to appear and eventually supercedes the degenerating female tissue. For the sea bass this transformation from female to male occurs at the age of about five years.

In protandrous hermaphrodism the testicular gonadal region predominates over ovarian tissue early in life and as the fish grows the reverse takes place. This form of hermaphrodism has been reported for members of Sparidae, Platycephalidae, and Gonostomatidae.

The majority of fishes are thought to be gonochoristic in which the species is either male or female. In the early development of the gonadal region the embryonic tissue is undifferentiated. If the indifferent gonad partially develops into an ovary like structure then about one-half of the individuals become males and the other half females. This is termed undifferentiated gonochorism. If the indifferent gonad differentiates directly into either ovary or testis, gonochorism is said to be differentiated.

Undifferentiated gonochorism is found in some lampreys and hagfishes. Small ammocoete larva possess indifferent gonads up to about 3.5 cm. Beyond this stage both male and female germ cells are found even though the organism is either a sperm producer or an egg producer. Among the teleosts, the eel *Anguilla anguilla,* the rainbow trout, *Salmo gairdneri irideus,* the herring *Clupea harengus* and the guppy *Lebistes reticulata* have been shown to be undifferentiated gonochorists.

Differentiated gonochorism has only been demonstrated for a few fishes and sexuality of this type is the most stable among fishes. The platyfish, *Xiphophorus maculatus,* and the Japanese medaka, *Oryzias latipes,* have been examined by the thousands without a single case of intersexuality (Yamamoto, 1969) . In both types of gonochorism sex differentiation is thought to be brought about by male and female inducing substances. These fishes that are known to be intersexes are thought to originate from the undifferentiated type while intersex is rarely found in differentiated gonochorists.

Polygenic Sex Determination

Until 1966 heteromorphic chromosomes associated with sex had not been reported. Based on genetic evidence, heterogametic forms were designated XX for female and XY (or XO) for the male if the male was heterogametic. If the female was heterogametic, they were commonly designated as WZ for the female and

ZZ for the male. The symbols to designate the various types of heterogamety in fishes over the years is perplexing. While no cytological evidence accompanied these studies, genetical experiments did provide information on the nature of sex-linked inheritance. Much of the genetical work was done on the live bearers of the family Poeciliidae. Specimens were crossed and various color markings, representative of a gene or gene group, were followed in the phenotypes of the offspring. In this way it was determined that certain genes, associated with the production of male or female markings, were located on either the X or Y chromosome. The authors of these studies reported that these genes were labile in nature and were not buffered against changing external conditions.

In most cases sex determination is considered to be polygenic rather than chromosomal. Chromosomes that carry genes that are dominant or epistatic to other genes for the determination of primary sex have been designated sex chromosomes. In addition, there are other genes distributed on other autosomes that also contribute to primary as well as secondary sex characteristics. Since heteromorphic chromosomes associated with sex in fishes have been the exception rather than the rule, sex determination was regarded as polygenic or polyfactorial.

In many instances the organism will have the genotype of one sex and the phenotype of the other. This has become evident in cases of sex reversal where mature species will have the gonadal structure of one sex and morphological characteristics associated with the other sex. There is doubt that true sex reversal involving primary sex characteristics occurs spontaneously. Partial success has been obtained surgically and through treatment with sex hormones. The gonads of the female Siamese fighting fish, *Betta splendens,* have been removed surgically and in rare instances a regenerated gonad with functional testes resulted. A number of claims have been made where reversal was chemically induced but only a few are considered to be true cases. *Oryzias latipes* is a differentiated gonochorist and the sex determining mechanism was determined by genetic crosses to be XX for female and XY for male. By treating the males with estrone and the females with methyl testosterone, it was possible to reverse the sex in each form so that mating was achieved in opposite fashion (Yamamoto, 1961). Other similar instances have been reported but while an organism may be transformed into a phenotypic and functional member of the opposite sex, the actual genetic constitution and distribution remains unchanged.

From what has been said thus far it would seem that sex development in fishes is unstable and not easily determined. Most fishes cannot be sexed externally. Some sexually dimorphic species exist with clear cut male or female secondary sex characteristics such as the gonopodium in viviparous cyprinodonts, and the urinogenital papilla in the female medaka. Secondary sex characteristics are usually transitory such as nuptial colorations that appear only during the breeding season. The only true way to determine primary sex is to autopsy the organism and decide if the gonads are testicular, ovarian, or intersex. If the organisms are young, and the gonads have not sufficiently developed, even autopsy will not confirm sex. If primary and secondary sex characteristics are this variable, and if the genes that produce these characteristics are polygenic, then it becomes impossible to cytologically ascertain sex chromosomes. Even if the majority of genes are on one chromosome pair, if this pair is not heteromorphic, the only way to determine if these were sex chromosomes would be through genetic crosses using secondary sex characteristics known to be associated with either the male or the female.

Heteromorphic Chromosomes and Heterogamety

Chen and Ebeling (1966) reported photokaryotypes of heteromorphic chromosomes in the deep sea fish *Bathylagus wesethi*. The organism had a $2n$ number of 36. The seventeen pairs of autosomes were classified into three groups. One large pair had satellites, seven pairs were submetacentric, nine pairs were metacentric, and the presumed pair of sex chromosomes consisted of a metacentric, which was the largest in the complement, and an acrocentric, which was the smallest in the complement. It was presumed that the large metacentric was the X chromosome and the small acrocentric was the Y chromosome. Testes squashes disclosed a connection between a large and small chromosome which was interpreted to be the X-Y bivalent. In some cases this bivalent was connected to a bivalented pair of metacentric chromosomes but was not considered to be the rule since distinct separations were found in some metaphase I spreads. In metaphase II, interpretations were made that some of the spreads had an X but no Y chromosome or vice versa. This was further supportive that the heteromorphic pair were sex chromosomes but still did not prove it. Ovarian metaphase analysis was not reported. This is understandable since meiotic studies from eggs are difficult to perform. The divisional stages usually are in an arrested condition and metaphases are rarely found.

Photokaryological evidence for female heterogamety was reported by Chen and Ebeling (1968) for the mosquitofish, *Gambusia affinis*. The somatic karyotype consisted of forty-eight chromosomes. The female had one metacentric, two submetacentrics and forty-five acrocentrics while the male had two submetacentrics and forty-six acrocentrics. It was presumed that the female was WZ and the male ZZ. Meiotic figures were obtained only from the testis. The Z-Z bivalent was undetectable because of its small size and heteropycnotic nature. Oogonial cells were examined for figures but no cells were observable after pachytene of prophase I. Metaphase II was seen in testis squashes but the small chromosomes were difficult to differentiate. Somatic chromosomes clearly contained heteromorphic pairs in both California and Texas populations. The author has also found this to be true with Alabama populations. Again, heterogamety in this species is only strongly suggested and evidence cannot be finalized until male and female meiotic metaphases are more fully assessed.

Chen (1969), reported heterogamety in twelve out of twenty-five deep sea fishes including *Bathylagus wesethi*. It was presumed that the heteromorphic pairs were of the XX-XY or XX-XO sex type. Chromosomal dimorphism in somatic cells was clearly defined for most of the organisms. Those that were considered to be X chromosomes were usually the largest in the complement and the Y chromosomes were the smallest. In some cases the Y chromosome could not be identified and in others an XO type mechanism was suggested. In all species the male was considered to be the heterogametic sex. Nine of the male species had a heteromorphic pair and was presumed to be X-Y *(Bathylagus wesethi, B. ochotensis, B. milleri, B. stilbius, Scopelengys tristis, Symbolophorus californiensis, Scopelogadus mizolepis bispinosus, Scopeloberyx robustus* and *Melamphaes parvus)* and three had only one heteromorphic chromosome and was presumed to be XO *(Sternoptyx diaphana, Lampanyctus ritteri,* and *Parvilux ingens).* In some cases, bivalent analysis reflected some evidence of heteromorphic pairing but all could not be examined with optimum clarity. When metaphase II figures were examined from specimens thought to be XO, some were interpreted to be with and others without the X chromosome and thus differed in total number by one chromosome.

Ebeling and Chen (1970) reviewed and summarized their previous findings in the light of studies with shallow water fishes. In the shallow water forms studied, five of thirty species had heteromorphic chromosome pairs in either the males or the females. In addition to *Gambusia affinis,* these consisted of two killifishes of the genus *Fundulus,* and two sticklebacks. The females of *Fundulus* had homomorphic submetacentric chromosomes and the males had heteromorphic pairs that were presumed to be XY chromosomes. One of the sticklebacks, *Apeltes quadracus* was female heterogametic and the other, *Gasterosteus wheatlandi* was presumed to be male heterogametic. Meiotic chromosomes were not evaluated in this particular study.

Multiple Sex Chromosomes

Uyeno and Miller (1971), reported a case of multiple sex chromosomes in an undescribed cyprinodontid killifish from Mexico in which the female had a $2n$ number of forty-eight and the male a $2n$ of forty-seven. Somatic chromosomes of both sexes contained one pair of metacentrics, eighteen pairs of submetacentric-subtelocentrics and four pairs of acrocentrics. In addition, the female had a pair of acrocentric chromosomes and the male a single large metacentric. As diakinesis there were twenty-two bivalents and one trivalent. This finding was interpreted to be a case of multiple sex chromosomes which paralleled that of the pygmy mouse, *Mus minutoides.* The female fish was presumed to have forty-four autosomes and $X_1X_1X_2X_2$ sex chromosomes while the male had forty-four autosomes and X_1X_2Y sex chromosomes. The gametes would then be twenty-two autosomes and X_1X_2 for the female and twenty-two autosomes and either X_1X_2 or the single Y for the male. This condition is thought to have originated from parents with forty-eight chromosomes in which the female had forty-six autosomes and X_1X_1 and the male forty-six autosomes and X_1Y. It was further suggested that all the sex chromosomes and some of the autosomes were acrocentric. Then in the male, the Y acrocentric centrically fused with an autosomal acrocentric to produce the large metacentric commonly found in male somatic spreads. During synapsis a trivalent configuration is formed as a result of the attachment of X_1 on the Y end and the odd autosome (now designated X_2) on the other end. By convention, the two autosomes of the female that are comparable to the rearranged autosomes of the male, are designated X_1 and X_2.

In 1972, the same authors (Uyeno and Miller, 1972) reported a second case of multiple sex chromosomes in an undescribed Mexican goodeid species in which the female had a diploid number of forty-two and the male forty-one. The female complement included six large metacentrics, two small metacentrics, two small submetacentrics or subtelocentrics, two medium sized subtelocentrics, and thirty acrocentrics of medium to small size. The male karyotype consisted of seven large metacentrics, two small metacentrics, two small submetacentrics or subtelocentrics, two medium sized subtelocentrics, and twenty-eight acrocentrics of medium to small size. Thus, the male had one more metacentric and two less acrocentrics than the female. In testis squashes, the diakinesis phase contained eighteen bivalents, three quadrivalents and a trivalent. Again, the male is thought to be X_1X_2Y and the female $X_1X_1X_2X_2$. This condition is thought to have arisen in the same way that was previously described for the killifish.

Concluding Remarks

From the work that has been done thus far, there is still no clear cut definition or trend established for sex chromosomes in fishes. A variety of sexuality mechanisms exists but most are thought to be bisexual or gonochoristic. Specific genes are located on chromosomes that control the synthesis of hormones which in turn determine primary sex. The chromosomes that carry these single genes, or blocks of genes, are designated sex chromosomes. If these chromosomes are structurally different, they are termed heteromorphic sex chromosomes. The organism with heteromorphic sex chromosomes is also heterogametic. Most fishes have homomorphic chromosomes and heterogamety is cytologically indeterminate. Sex types in homomorphic forms can be determined through genetical crosses and through studies of sex reversal. From recent studies, there are more fishes with heteromorphic chromosomes than was previously thought. Cytologically, sex determining types have been reported that include XX-XY, XX-XO, WZ-ZZ and multiple sex chromosomes.

It will be evidenced from the next chapter that chromosomes from more than five hundred fishes have been studied to date. As more fishes are studied cytologically with the techniques that are available, such as banding methods, autoradiography, bivalent analysis, and Metaphase II analysis, there will be more answers to important questions concerning our understanding of sex determination in fishes and the roles of sex chromosomes in development.

CHAPTER 5

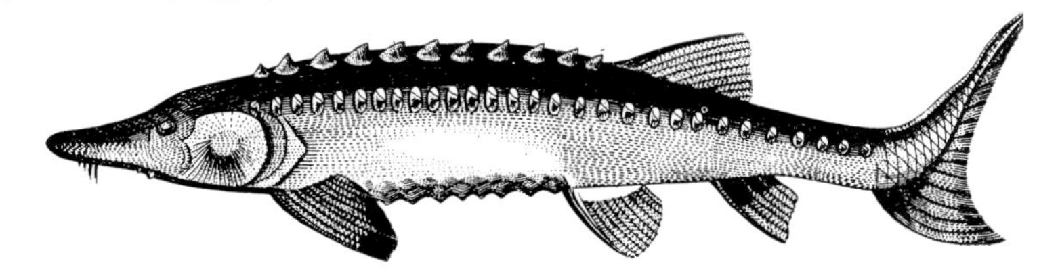

A CHECK-LIST OF FISH CHROMOSOMES

Of the twenty thousand known species of fishes, chromosomes have been reported for less than six hundred species. The accompanying check list has 616 entries representing 481 species. Of this number, the chromosome types are given for 285 species. Specimens are represented for all but sixteen of the forty-one orders of fishes. All the orders are listed in the check list and those not having representatives that have been studied are noted in parenthesis.

The chromosome numbers and types are listed phylogenetically. The species are listed alphabetically under each family for convenience. The classification system is that given in Chapter 1. The Agnathans and Chondrichthyes are classified according to Berg (1947) as given by Lagler, *et al.* (1962). The major classification of the Osteichthyes is that of Grassé (1957) and the teleost are classified according to Greenwood, *et al.* (1967).

The chromosome numbers and types came from a number of sources. The majority were compiled from individual reprints. All those denoting chromosome types were obtained in this way. Others were obtained from listings reported by Makino (1957), Gyldenholm and Scheel (1971), Nogusa (1960) and Roberts, (1967). Information is given for the diploid chromosome number, haploid number and the chromosome type for the diploid complement. The terms metacentric, submetacentric, and acrocentric are given for this purpose. It would have been preferred that the term subtelocentric also be used but the number of authors that used this term were so few that it was not significant enough to justify its placement all the way through the listing. All terms were taken directly from reprint descriptions. No attempt was made to interpret the author's karyotypes. It is realized that the listing as presented does not include every chromosome number or type that has ever been reported. To the investigators whose works were unknowingly omitted, the author stands apologetic.

AMPHIOXI

CLASSIFICATION	CHROMOSOME NUMBER		CHROMOSOME TYPE			REFERENCES
	2*n*	*n*	M	SM	A	
Amphioxiformes						
Branchiostomidae						
Amphioxus lanceolatus		10				1,2,3
” ”	24	12				4,5,6
Branchiostoma belcheri	32	16	0	0	32	7
Branchiostoma floridae	38	19				8

AGNATHA

CLASSIFICATION	CHROMOSOME NUMBER		CHROMOSOME TYPE			REFERENCES
	2*n*	*n*	M	SM	A	
Myxiniformes						
Eptatretidae						
Eptatretus burgeri	48	24	0	0	48	7,9
Eptatretus okinoseanus	46	23	0	0	46	7,9
Eptatretus stoutii	48	24				10
Myxinidae						
Myxine glutinosa	50					11
” ”	52	26				12,13
Petromyzontiformes						
Petromyzontidae						
Bdelostoma burgeri	48					14
Entosphenus reissneri	94-96		0	0	94-96	7
” ”	165-174					15
Ichthyomyzon gagei	164					16
Lampetra aepyptera	164					17

AGNATHA (Continued)

CLASSIFICATION	CHROMOSOME NUMBER		CHROMOSOME TYPE			REFERENCES
	2*n*	*n*	M	SM	A	
Mordacia mordax	76					18
Mordacia praecox	76	38				19
Petromyzon marinus	168					20

CHONDRICHTHYES

CLASSIFICATION	CHROMOSOME NUMBER		CHROMOSOME TYPE			REFERENCES
	2*n*	*n*	M	SM	A	
Squaliformes						
Scyliorhynidae						
Pristiurus melanostomus	30-50					21
Scyllium canicula	30-50					22
” ”	24	12				23,24,25
” ”	14-24					26
” ”	17-19					27
Carchariidae						
Mustelus manazo	72	36	—12—		60	7
Squalidae						
Scylliorhinus catula		31				28,29
Spinax niger	60-70					30
Squalus acanthias	62	31				31
Squalus suck'eyi	62	31				32
Rajiformes						
Rajidae						
Raja batis	98	49	0-6		92-98	31
Raja clauata	98	49	0-6		92-98	31

CHONDRICHTHYES (Continued)

CLASSIFICATION	CHROMOSOME NUMBER		CHROMOSOME TYPE			REFERENCES
	2*n*	*n*	M	SM	A	
Raja macrorhynchus	24	12				24,25
Raja maculata	24	12				24,25
Raja meerdervoortii	104	52				32
Raja radiata	98	49	0-6		92-98	31
Dasyatidae						
Dasyatis akajei	84	42	—20—		64	7
Torpediniformes						
Torpedidae						
Torpedo ocellata		30-50			33	33
Chimaeriformes						
Chimaeridae	58	(34 microchromosomes)			34	34
Hydrolagus colliei						

DIPNEUSTI

CLASSIFICATION	CHROMOSOME NUMBER		CHROMOSOME TYPE			REFERENCES
	2*n*	*n*	M	SM	A	
Lepidosireniformes						
Lepidosirenidae						
Epiceratodus forsteri	32-38		32-38	0	0	35
Lepidosiren paradoxa	38		38	0	0	36
" "	36		34	0	0	37
" "	38	19	38	0	0	38,39
Protopterus annectens	34	17				35

BRACHIOPTERYGII

CLASSIFICATION	CHROMOSOME NUMBER		CHROMOSOME TYPE			REFERENCES
	2n	n	M	SM	A	
Polypteriformes						
Polypteridae						
Calamoichthys calabaricus	36		30	6	0	40
Polypterus palmas	36		24	12	0	40

CHONDROSTEI

CLASSIFICATION	CHROMOSOME NUMBER		CHROMOSOME TYPE			REFERENCES
	2n	n	M	SM	A	
Acipenseriformes						
Acipenseridae						
Scaphirhynchus platorhynchus	112 (48 dot-like)		50	0	14	34

HOLOSTEI

CLASSIFICATION	CHROMOSOME NUMBER		CHROMOSOME TYPE			REFERENCES
	2n	n	M	SM	A	
Amiiformes						
Amiidae						
Amia calva	46		20	0	26	34
Lepisosteiformes						
Lepisosteidae						
Lepisosteus productus	68		28	0	14	34

TELEOSTEI

CLASSIFICATION	CHROMOSOME NUMBER		CHROMOSOME TYPE			REFERENCES
	2n	n	M	SM	A	
El piformes (None)						
Anguilliformes						
Moringuidae						
Moringua linearis	50					137
Anguillidae						
Anguilla anguilla	38		14	10	14	41
" "	38					42
Anguilla japonica	38					42
Anguilla rostrata	38					42
Anguilla vulgaris	36					43
Echelidae						
Echelus uropterus	50	25	0	4	46	7
Muraenidae						
Gymnothorax kidako	42	21	12	0	30	7
Muraena paradalis	40	20	14	0	26	7
Notacanthiformes (None)						
Clupeiformes						
Clupeidae						
Alosa pseudoharengus	48	24	0	0	48	44
Clupea harengus	52					45
Engraulidae						
Englaulis japonicus	48	24	0	0	48	7
Englaulis mordax	48		0	0	48	46

TELEOSTEI (Continued)

CLASSIFICATION	CHROMOSOME NUMBER		CHROMOSOME TYPE			REFERENCES
	2*n*	*n*	M	SM	A	
Osteoglossiformes						
Nototeridae						
Notopterus chitala	48	24	12	0	36	47
Notopterus notopterus	48	24	12	0	36	47
" "	42	21	0	0	42	48
Mormyriformes (None)						
Salmoniformes						
Salmonidae						
Oncorhynchus gorbuscha	52		—52—		0	49
Oncorhynchus keta	74					50
" "	74		—26—		48	51
" "	74		—28—		46	49
" "		50				7
Oncorhynchus kisutch	60		—52—		8	49
Oncorhynchus masov		50				7
Oncorhynchus nerka		54				7
" "	58		—44—		14	51
" "	56		—46—		10	49
O. nerka x *O. keta*	66		—35—		31	51
Oncorhynchus rhodurus		50				7
Oncorhynchus tshawytsch	68		—36—		32	49
Salmo alpinus	80		—16—		64	52
" "	80	40	20	0	60	53
Salmo carpio	96					54
Salmo clarki	64		—42—		22	49

TELEOSTEI (Continued)

CLASSIFICATION	CHROMOSOME NUMBER		CHROMOSOME TYPE			REFERENCES
	2*n*	*n*	M	SM	A	
Salmo clarki lewisi	64					55
Salmo fario		12				56
Salmo fontinalis	84					52
Salmo gairdneri	60		—44—		16	55
” ”	60		44	0	16	57
Salmo irideus	104	52	—14—		90	7
” ”	60 ±		43	0	18	36
” ”	60					58
” ”	68		36	0	32	46
” ”	58-65					59
Salmo marmoratus	84					54
Salmo salar	54-56		—18—		36	60
” ”	57					63
” ”	60					61
” ”	56		—16—		40	62
” ”	60		—12—		48	52
” ”	58		—16—		42	64
” ”	58	29	—16—		42	65
Salmo salar x *S. trutta*	70	32-38				52
S. salar x *Salvelinus fontinalis*	70					61
Salmo trutta	80		—18—		62	52
” ”	80	40	20	0	60	53
Salmo (Trutta) *lacustris*		24				66
” ” ”	70					54
S. trutta x *S. salar*		20-30				52

TELEOSTEI (Continued)

CLASSIFICATION	CHROMOSOME NUMBER		CHROMOSOME TYPE			REFERENCES
	2*n*	*n*	M	SM	A	
Salvelinus alpinus	80		—16—		64	52
Salvelinus fontinalis	84		—16—		68	52
” ”		50				7
” ”	84					67
Salvelinus namaycush	84		—16—		68	68
Coregonus albula	80		—16—		64	52
” ”	80	40	16	0	64	69
Coregonus artedii	80		16	10	54	70
Coregonus asperi	36					71
Coregonus asperi marenoides	36					71
Coregonus clupeaformis	80		20	8	52	70
Coregonus exigus	72					71
Coregonus exigus albellus	70 ±					71
Coregonus nasus	96				96	72
” ”	80	80	12	0	68	73
Coregonus hoyi	80		20	8	52	70
Coregonus lavaretus	80	40	12-18	0	62-68	73
” ”	80		—16—		64	52
” ”	96				96	72
” ” *baunti*	80		—16—		64	74
” ” *baeri*	80					61
C. l. baeri x *S. fontinalis*	80					61
Coregonus oxyrhynchus	96				96	72
Coregonus peled	80		—12—		68	74
Coregonus peled	80	40	12-18	—	62-68	73

TELEOSTEI (Continued)

CLASSIFICATION	CHROMOSOME NUMBER		CHROMOSOME TYPE			REFERENCES
	2*n*	*n*	M	SM	A	
Coregonus pidschian	80	40	12-18	—	62-68	73
” ”	96				96	72
Coregonus reighardi	80		12	12	56	70
Coregonus schizii	72					71
Coregonus wartmanni	78 ± 3					75
Coregonus wartmanni coeruleus	70 ± 2					71
Coregonus zenithicus	80		10	8	62	70
Prosopium coulteri	82		18	0	64	70
Prosopium cylindraceum	78		22	0	56	70
Thymallus thymallus	102	51	68	0	34	69
” ”	130		—28—		74	52
Osmoridae						
Osmerus eperlanus	58		10			52
” ”	54	27	16	0	38	76
Bathylagidae						
Bathylagus milleri	60					77
Bathylagus ochotensis	54					77
Bathylagus stilbius	64					77
Bathylagus wesethi	36	18				77
” ”	36	18	19	14	1	78
Salangidae						
Salangichthys microdon		28				7
Galaxiidae						
Brachygalaxis bullocki	38		10	16	12	94
Galaxis maculatus	22		8	12	2	94

TELEOSTEI (Continued)

CLASSIFICATION	CHROMOSOME NUMBER		CHROMOSOME TYPE			REFERENCES
	2*n*	*n*	M	SM	A	
Galaxis platel	30		1-2	16-18	10-12	94
Esociade						
Esox americanus	50	25	0	0	50	79
Esox americanus vermiculatus	50				50	79
Esox americanus americanus	50	25	0	0	50	79
Esox lucius	50		0	0	50	79
" "	48					80
" "	18					52
" "	50	25				81
E. lucius x *masquinony*	50		0	0	50	79
E. lucius x *E. reicherti*	50	25	0	0	50	79
Esox masquinony	50	25	0	0	50	79
Esox niger	50		0	0	50	79
Umbra limi	20-22	11				82
Sternoptychidae						
Sternoptyx diaphana	35	17 (Male) & 18 (Female)				77
Myctophidae						
Lampanyctus ritteri (Male)	47	23				77
" " (Female)	48	24				77
Parvilux ingens (Male)	49	24				77
" " (Female)	50	25				77
Symbolophorus californiensis	48	24				77
Neoscopelidae						
Scopelengys tristis	48	24				77
Cetomimiformes (None)						

TELEOSTEI (Continued)

CLASSIFICATION	CHROMOSOME NUMBER		CHROMOSOME TYPE			REFERENCES
	$2n$	n	M	SM	A	
Ctenothrissiformes (None)						
Cypriniformes						
Characinidae						
Alestes longipinis		24				83
Anoptichthys jordani		24				83
Astyanax mexicanus		24				83
Chalceus macrolepidotus	54		—32—		22	84
Cheirodon axelrodi	52					85
" "		24				83
Hasemania marginata		24				83
Hemigrammus ocellifer		24				83
Hemigrammus pulcher		24				83
Hemigrammus schmardae	52					85
Hyphessobrycon flammeus		24				83
Hyphessobrycon gracilis	48	24				83
Hyphessobrycon griemi		24				83
Hyphessobrycon haraldschultzi		24				83
Hyphessobrycon pulchripinnis		24				83
Hyphessobrycon serpae		24				83
Hyphessobrycon simulans		25				86
Hyphessobrycon stictus	52					85
Megalamphodus megalopterus		24				83
Micralestes interruptus		24				83
Moenkhausia pittieri		24				83
Moenkhausia sanctae-filomenae		24				83

TELEOSTEI (Continued)

CLASSIFICATION	CHROMOSOME NUMBER		CHROMOSOME TYPE			REFERENCES
	2*n*	*n*	M	SM	A	
Paracheirodon innesi	32					85
” ”	36					87
Pristella riddlei		24				83
Serrassalmus hollandi	64		—30—		34	84
Apteronotidae						
Apteronotus albifrons	24		14	2	8	88
Cyprinidae						
Abbottina rivularis	50		6	44	0	89
” ”	50	25	6	44	0	7
Abramis brama		26				83
” ”	52					90
Acheilognathus cyanostigma	44	22	4	0	40	7
Acheilognathus lanceolatus	50		4	0	46	91
Acheilognathus limbata	44	22	4	0	40	7
Acheilognathus rhombea	44-48	22-24	0-44	0	40-48	7
Amblypharyngodon mola	52	26	8	0	44	92
Barbus conchonius		24				83
Barbus fasciatus	52		—30—		22	93
Barbus oligolepis		24				83
Barbus semifaciolatus	52	26	4	0	48	7
Barbus tetrazona	50		—34—		16	93
” ”	50		34	6	10	7
” ”		24				83
Barbus titteya		24				83
Barbus viviparus		24				83

TELEOSTEI (Continued)

CLASSIFICATION	CHROMOSOME NUMBER		CHROMOSOME TYPE			REFERENCES
	2n	n	M	SM	A	
Brachydanio albolineatus		24				83
Brachydanio frankei		24				83
Brachydanio rerio		24				83
Carassius auratus	94	47				95,96,97
" "	104		—46—		58	83
" "	102 ±		64		37	36
" "	104		20	—84—		41
" "	100		20	40	40	58
Carassius carassius	94	47				97
" "	94					98
Chela bacaila	52	26	0	0	52	92
Chrosomus eos	50					99
C. oes x *C. neogaeus*	50					99
Crossochelius punjabensis	48	24	12	0	36	92
Ctenopharyngodon idellus	48	24	0	0	48	7
Cyprinus carpio	104	52				91,95
" "	104	52	—46—		58	93
" "	100		12	36	52	100
" "	104		46	18	40	101
C. carpio x *Carassius carassius*	99					97
C. auratus x *Carrassius carassius*		47				97
Exoglossum maxillingua	48					99
Gnathopogon elongatus caerulescens	50	25	0	0	50	7
" " "	50	25	0	0	50	7
Gobio gobio		24				83

TELEOSTEI (Continued)

CLASSIFICATION	CHROMOSOME NUMBER		CHROMOSOME TYPE			REFERENCES
	2*n*	*n*	M	SM	A	
Hemibarbus longirostris	50	25	0	0	50	7
Hemigrammocypris rasborella	48	24	0	0	48	7
Hybopsis plumbea	50					99
Ishikavia steenackeri	48	24	2	0	46	7
Labeo calbasu	54	27	0	0	54	92
Labeo chrysophekadion	50		—14—		36	84
Labeo devo	54	27	0	0	54	102
Labeo gonius	54	27	0	2	52	103
" "	54	27	0	0	54	102
Misgurnus anguillicaudatus	52					98
Moroco percnurus	54	27	8	0	46	7
Moroco steindachneri	54	27	8	0	46	7
Notemigonus crysoleucas	50		16	24	10	104
Notropis callistius	50					105
Notropis lutrensis	50		10	32	8	104
Notropis stilbius	50					105
Opsariichthys unicrostris	66	33	0	0	66	7
Oxygaster bacaila	50	25	0	0	50	103
Phoxinus percnurus sachalinensis		27				106
Pimephales notatus	52					99
Pseudogobio esocinus	52	26	0	0	52	7
Pseudorasbora parva	50		4	0	46	91
Ptychochelius lucius	50					107
Pungtungia herzi	50		4	0	46	7
Pungtungia hilgendorfi	50					106

TELEOSTEI (Continued)

CLASSIFICATION	CHROMOSOME NUMBER		CHROMOSOME TYPE			REFERENCES
	2*n*	*n*	M	SM	A	
Puntius conchonius	50	25	0	0	50	92
Puntius sophore	50	25	0	0	50	92
Puntius stigma	50	25	0	0	50	92
" "	48	24	0	0	48	103
Puntius ticto ticto	50	25	0	0	50	103
Rasbora heteromorpha		24				83
Rasbora trilineata		24				83
Sarcochielichthys variegatus	50	25	0	0	50	7
Sardinius erythrophthalmus	52					90
Sardinius erythrophthalmus	52					83
" "	46		12	—34—		41
Semotilus atromaculatus	52					99
Semotilus corporalis	52					99
Semotilius margarita	50					99
Tinca tinca	46					41
Tribolodon hakuensis	50		6	44	0	91
Typhlogarra widdowsoni		24				83
Zacco platypus	48	24	0	0	48	7
Zacco temmincki	50		4	0	46	7
Cobitidae						
Acanthophthalmus khulli	50		—14—		36	93
Barbatula oreas	48		4	0	44	97
Barbatula toni	50	25	—16—		34	108
Botia macracantha	98		—28—		70	93
Cobitis biwae	54	27	4	0	50	7
" "	96		—58—		38	108

TELEOSTEI (Continued)

CLASSIFICATION	CHROMOSOME NUMBER		CHROMOSOME TYPE			REFERENCES
	2n	n	M	SM	A	
Cobitis delicata	50	25	—18—		32	108
Lefua echigonia	50	25	0	0	50	7
Lefua nikkonis	50	25	—12—		38	108
Misgurnus anguillicaudatus	52				52	97
” ”	50	25	—14—		36	108
” ”	52				52	98
Catostomidae						
Catostomus discobolus	100					107
Erimyzon sucetta	100					107
Moxostoma duquesnei	100					107
Siluriformes						
Ictarulidae						
Ictarulus punctatus	56		—16—		40	84
Bagridae						
Mystus seenghala	50	25	30	0	20	109
Mystus tengara	54	27	10	0	44	110
Mystus vittatus	50					109
Pelteobagrus nudiceps	56	28	0	0	56	7
Rita rita	54	27	16	0	38	110
Siluridae						
Ompok bimaculatus	40	20	4	0	36	110
Parasiluris asotus	58	29	0	0	58	7
Wallago attu	86	43	0	0	86	110
Schilbeidae						
Clupisoma garua		33				110

TELEOSTEI (Continued)

CLASSIFICATION	CHROMOSOME NUMBER		CHROMOSOME TYPE			REFERENCES
	2*n*	*n*	M	SM	A	
Clariidae						
Clarius batrachus	52	26	0	0	52	110
Loricariidae						
Hypostomus plecostomus	54		—24—		30	84
Loricaria parva		24				83
Heteropneustidae						
Heteropneustes fossilis	58	29	0	0	58	110
Percopsiformes (None)						
Batrachoidiformes (None)						
Gobiesociiformes (None)						
Lophiiformes (None)						
Gadiformes (None)						
Atheriniformes						
Belonidae						
Xenetodon cancila	50	25	0	0	50	103
Exocoetidae						
Dermogenys pusillus		24				83
Cyprinodontidae						
Aphanius cypris	48	24				111
Aphanius dispar		24				111
Aphanius fasciatus		24				111
Aphanius iberus		24				111
Aphanius sophiae		24				111
Aphyoplatys duboisi	48	24				112,113
Aphyosemion ahli		18				113

TELEOSTEI (Continued)

CLASSIFICATION	CHROMOSOME NUMBER		CHROMOSOME TYPE			REFERENCES
	2n	n	M	SM	A	
Aphyosemion arnoldi	38	19				113
” ”		20				83
Aphyosemion australe	30	15				112
Aphyosemion bivittatum	40	20				112
” ”	26-40	13-20				113
Aphyosemion bivittatum hollyi		20				83
Aphyosemion bualanum	38-40	19-20				113
Aphyosemion calliurum	32	16				112,113
Aphyosemion calliurum ahli	40	20				83
Aphyosemion cameronense	34	17				113
Aphyosemion celiae	20	10				114
Aphyosemion christyi	18-30	9-15				113
Aphyosemion cinnamomeum	40	20				113
Aphyosemion exiguum	36	18				113
Aphyosemion filamentosum	36	18				112,113
Aphyosemion franzwerneri	22	11				114
Aphyosemion gardneri	36-40	18-20				113
Aphyosemion gulare	32	16				112,113
Aphyosemion labarrei	28	14				113
Aphyosemion louessense	20	10				112,113
Aphyosemion lujare	40	20				113
Aphyosemion ndianum	40	20				113
Aphyosemion obscurum	34	17				113
Aphyosemion santa-isabellae	40	20				115
Aphyosemion sjoestedti	40	20				113
” ”		20				83

TELEOSTEI (Continued)

CLASSIFICATION	CHROMOSOME NUMBER		CHROMOSOME TYPE			REFERENCES
	2n	n	M	SM	A	
Aphyosemion walkeri	36	18				113
Aplocheilichthys flavipinnis		24				83
Aplocheilichthys katangae		24				83
Aplocheilichthys spilauchenia		24				83
Aplocheilus blockii	48	24				112,113
Aplochelius dayi	48	24				112,113
Aplocheilus lineatus	50	25				112,113
Aplochelius panchax	36	18				112,113
Austrofundulus dolichopterus	44	22				116
Cynolebias bellotti		24				83
Cynolebias nigripinnis		24				83,117
Cynolebias whitei		23				117
" "		24				83
Cynopoecilus ladigesi		24				83
Cynopoecilus melanotaenia		22				117
Cyprinodon macularius		24				83
Epiplatys annulatus	50	25				112,113
Epiplatys barmoiensis		17				113
Epiplatys bifasciatus	40	20				112,113
Epiplatys chaperi	50	25				112,113
" "		24				83
Epiplatys dageti	50	25				112,113
Epiplatys esekanus	42	21				113
Epiplatys fasciolatus	38	19				112,113
Epiplatys grahami	46	23				113

TELEOSTEI (Continued)

CLASSIFICATION	CHROMOSOME NUMBER		CHROMOSOME TYPE			REFERENCES
	2n	n	M	SM	A	
Epiplatys sangmelinensis		24				113
Epiplatys sexfasciatus	48	24				83,112,113
Epiplatys spilargyreius	34	17				112,113
Fundulus chrysotus	34		14	0	20	118
Fundulus cingulatus	46		2	0	44	118
Fundulus confluentus	48		0	0	48	118
Fundulus diaphanus (male)	48	24	1 (2 SAT)*	3	42	119
Fundulus diaphanus (female)	48		0 (2 SAT)*	4	42	119
Fundulus diaphanus	48		0	4	44	118
" "	48					120
Fundulus grandis	48		0	2	46	118
Fundulus heteroclitus	36					121
" "	45					122
" "	48					120,124
" "		24				83
Fundulus heteroclitus	48	24	0 (2 SAT)*		46	119
" "	48		0	0	48	118
Fundulus kansae	48		0	0	48	118
Fundulus lineolatus	46		2	0	44	118
Fundulus majalis	48		0	2	46	118
Fundulus majalis	48	24	0	2	46	119
" "	48					120,123
Fundulus notatus	40	20	10	0	30	124
" "	40		10	2	28	118

*Satellite Chromosomes

TELEOSTEI (Continued)

CLASSIFICATION	CHROMOSOME NUMBER		CHROMOSOME TYPE			REFERENCES
	2n	n	M	SM	A	
Fundulus notti	48					105
” ”	46		2	0	44	118
Fundulus olivaceous	48		2	0	46	124
” ”	48		2	2	44	118
Fundulus parvipinnis (female)	48		0	2	46	119
” ” (male)	48	24	0	1	47	119
Fundulus parvipinnis	48		0	0	48	118
” ”	48					120
Fundulus pulvereus	48		0	0	48	118
Fundulus rathbuni	48		0	2	48	118
Fundulus sciadicus	44		4	2	38	118
Fundulus seminolis	48		0	0	48	118
Fundulus similis	48		0	26	46	118
Fundulus stellifer	48					105
Fundulus waccamensis	48		0	4	42	118
Heterandria formosa	46	23				126
Jordanella floridae		24				83
Kosswigichthys asquamatus		24				111
Lebistes reticulatus	46	23				126,127,128,130,131
” ”		23				129
Limi vittata	46	23				126
Limi cnadofasciata tricolor	46	23				126
Mollienisia sphenops	36					132
” ”	46	23				126
Mollienisia sphenops melanistica	38	19				126

TELEOSTEI (Continued)

CLASSIFICATION	CHROMOSOME NUMBER		CHROMOSOME TYPE			REFERENCES
	2n	n	M	SM	A	
M. sph. mel. x *M. sph.*	42	21				126
M. sph. x *M. velifera*	46	23				126
Mollienisia velifera	46	23				126
Notobranchius guentheri		19				112,113
Notobranchius mayeri		19				83
Notobranchius orthonotus		18				113
Notobranchius palmaquisti		19				83
Notobranchius rachovii	16	8				83
" "	18	9				112,113
Orestias agassii		24				133
Orestias luteus		24				133
Oryzias latipes		22-24				134
Oryzias latipes	48	24				130,131,135
Oryzias melastigma	48		24	10	14	136
Pachypanchax playfairi	48	24				112,113
Procatopus nigromarginatus		24				117
Procatopus nototaenia		24				83,117
Procatopus roseipinnis		24				117
Procatopus similis		24				117
Pterolebias longipinnis		10				83
Pterolebias peruensis		27				83,117
Rivulus cylindraceus		24				83
Rivulus santensis	48	24				126
Roloffia bertholdi	42	21				112,113
Roloffia geryi	40	20				113

TELEOSTEI (Continued)

CLASSIFICATION	CHROMOSOME NUMBER		CHROMOSOME TYPE			REFERENCES
	2*n*	*n*	M	SM	A	
Roloffia guineensis	38	19				112,113
Roloffia occidentalis	46	23				113
Roloffia rolofii	42	21				112,113
Poecilliidae						
Belonesox belizanus		24				83
Gambusia affinis (female)	48	24	1	2	45	138
" " (male)	48	24	0	2	46	138
Gambusia affnis holbrookii	48	24			48	139
" " "	36	18				125
Gambusia gaigei (female)	48		1	1	46	141
" " (male)	48		0	0	48	141
Gambusia hurtadoi (female)	48		1	0	47	141
" " (male)	48		0	0	48	141
Gambusia marshi	42	21	0	0	42	141
Gambusia nobilis (female)	48		1	0	47	141
" " (male)	48		0	0	48	141
Gambusia regani	48	24	0	0	48	141
Gambusia vittata	48	24	0	0	48	141
Heterandia formosa		24				83
Mollienesia formosa	46		0	0	46	141
Phallichthys amates		24				83
Phallichthys pitteri	46	23				142
Phalloceros caudomaculatus	46	23				142
Phalloceros caudomaculatus reticulatus	46	23				142
Platypoecilus couchiana		24				143,144,145

TELEOSTEI (Continued)

CLASSIFICATION	CHROMOSOME NUMBER		CHROMOSOME TYPE			REFERENCES
	2n	n	M	SM	A	
Platypoecilus maculatus		24				143,144,145
Platypoecilus variatus		25				142,145
Platypoecilus xiphidium		24				145
P. couchiana x *P. maculatus*		24				145
P. variatus x *P. maculatus*		24-25				145
P. xiphidium x *P. maculatus*		24				145
Poecilia caudofasciata	46	23				142
Poecilia formosa	46, 69					146
Poecilia latipinna	46	23				146
" "		24				83
Poecilia latipunctata		23				146
Poecilia melanogaster		24				83
Poecilia mexicana	46	23				146
Poecilia reticulata	46	23				83,148
Poecilia sphenops	46	23				142,146
" "		24				83
Poecilia velifera	46	23				83,142,146
Poeciliopsis latidens	48					148
Poeciliopsis lucida	48, 72					148
Poeciliopsis monacha	48, 72					149
Rivulus santensis	48	24				142
Xiphophorus helleri	48	24				142,150
Xiphophorus helleri	48		0	0	48	36
" "		24				145
Xiphophorus maculatus	48	24				142,150

TELEOSTEI (Continued)

CLASSIFICATION	CHROMOSOME NUMBER		CHROMOSOME TYPE			REFERENCES
	2n	n	M	SM	A	
Xiphophorus montezumae	48	24				87
" "		24				145
Xiphophorus xiphidium		24				87
X. helleri x *P. maculatus*		24				145
Atherinidae						
Melanotaenia maccullochi		24				83
Menidia notata	36					151
Beryciformes						
Malamphaidae						
Melamphaes parvus	50	25				77
Scopeloberyx robustus	42	21				77
Scopelogadus mixolepis bispinosus	46					77
Zeiformes (None)						
Lampridiformes (None)						
Gasterosteiformes						
Gasterosteidae						
Apeltes quadracus	46					120
" " (Male)	46	23	12	20	14	152
" " (Female)	46	23	12	19	15	152
Culaea inconstans	46	23	—	8	38	152
Gasterosteus aculeatus	42	21	6	6	30	152
" "		21				83
" " *aculeatus*	42		—16—		26	153,154
Gasterosteus aculeatus microcephalus	42		—16—		26	153,154
Gasterosteus wheatlandi	42		0	0	42	120

TELEOSTEI (Continued)

CLASSIFICATION	CHROMOSOME NUMBER		CHROMOSOME TYPE			REFERENCES
	2n	n	M	SM	A	
” ”	42	21	4	6	32	152
Pungitius pungitius	42	21	16	12	14	152
Pungitius pungitius		21				83,95,155
” ”	42		—12—		30	154
Pungitius sinensis	42		—8—		34	153,154
Pungitius tymensis	42	21				95,155
” ”	42		—12—		30	154
Channiformes						
Channidae						
Channa marulius	40	20	10	0	30	110
Channa punctatus	34	17	16	0	18	110
Channa striatus	40	20	10	0	30	110
Synbranchiformes (None)						
Scorpaehiformes						
Synanceidae						
Inimicus japonicus	50	25	0	0	50	7
Hexagrammidae						
Hexagrammos octogrammus	24					157
Hexagrammos otakii	48	24	0	0	48	7
Cottidae						
Cottus baridii	36-38	18				158
Cottus gobio	52		6	0	46	159
Cottus poecilopus	48		8	0	40	159
Cottus pollux	48	24	0	0	48	7,11
Misgurnus fossilis		24				83

TELEOSTEI (Continued)

CLASSIFICATION	CHROMOSOME NUMBER		CHROMOSOME TYPE			REFERENCES
	$2n$	n	M	SM	A	
Dactylopteriformes (None)						
Pegasiformes (None)						
Perciformes						
Centropomidae						
Ambassis nama	50	25	0	0	50	110
Chanda ragna		22				83
Serranidae						
Coreoperca kawamebari	48	24	0	0	48	7
Theraponidae						
Therapon jabua	48					161
Therapon puta	48					161
Centrarchidae						
Acanthrarchus pomotis	48		0	0	48	162
Ambloplites repestris	48	24	0	0	48	162
Centrarchus macropterus	48	24	0	0	48	162
Chaenobryttus gulosus	48	24	0	0	48	162
Elassoma evergladei		24				83
Elassoma zonatum	48		0	2	46	162
Enneachanthus chaetodon	48	24	0	0	48	162
Enneachanthus gloriosus	48	24	0	0	48	162
Enneachanthus obesus	48	24	0	0	48	162
Lepomis auritus	48	24	0	0	48	162
Lepomis cyanellus	46-48	23-24	0	0	46-48	36,162
" "	46-48		0-2	0	44-48	163
" "	48		0	0	48	46

TELEOSTEI (Continued)

CLASSIFICATION	CHROMOSOME NUMBER		CHROMOSOME TYPE			REFERENCES
	2n	n	M	SM	A	
Lepomis gibbosas	48	24	0	0	48	162
Lepomis hymilis	46	23	—2—		44	162
Lepomis macrochirus	48	24	0	0	48	162
” ”	44					164
Lepomis marginatus	48		0	0	48	164
Lepomis megalotis	48	24	0	0	48	164
Lepomis microlophus	48	24	0	0	48	164
Micropterus dolomieni	46	23	0	2	44	162
Micropterus salmoides	46	23	—2—		44	162
Pomoxis annularis	48	24	0	0	48	162
Percidae						
Acernia cernua	48	24				81
Lucioperca lucioperca	48	24				81
Perca fluviatilis	48	24				81
Sillaginidae						
Sillago sihama	48	24	0	0	48	7
Pomadasyidae						
Haemulon sciurus	46		2	0	44	165
Monodacytlidae						
Monodactylus argenteus		24				83
Nandidae						
Badis badis		24				83
Nandus nandus	48	24				166
Cichlidae						
Apistrogramma pertense		24				83

TELEOSTEI (Continued)

CLASSIFICATION	CHROMOSOME NUMBER		CHROMOSOME TYPE			REFERENCES
	2n	n	M	SM	A	
Apistrogramma ramireze		24				83
Cichlasoma severum		24				83
Haplochromis multicolor		22				83
Hemichromis bimaculatus		22				83
Lamprologus leleupi		24				83
Pelmatochromos kribensis		24				83
Pterophyllum eimekei		24				83
Symphysodon aequifascata	60		44	0	16	36
Tilapia grahami		24				83
Tilapia mossambica	44	22				167
Mugilidae						
Mugil corsula	48	24	0	0	48	110
Labridae						
Ctenolabrus adspersus	38-48					122
Pseudolabrus japonicus	46		2	0	44	7
Bathymasteridae						
Ronquilus jordani		26				168
Mugiloididae						
Cilias pulchella	48	24	48	0	0	7
Pholididae						
Pholis pictus	46					50
Callionymidae						
Callionymus richardsoni	38	19	0	0	38	7
Gobiidae						
Acanthogobius flavimanus	44	22	0	0	44	7

TELEOSTEI (Continued)

CLASSIFICATION	CHROMOSOME NUMBER		CHROMOSOME TYPE			REFERENCES
	2*n*	*n*	M	SM	A	
Boleophthalmus pectinirostris	46	23	0	0	46	7
Brachygobius nunus		24				83
Chaenogobius urotaenia	44	22	0	0	44	7
Gillichthys mirabilis	44	22	0	12	32	169
Glossogobius giuris	46	23	0	0	46	166
Gobius abei	46	23	0	2	46	7
Gobius similis	44	22	0	0	44	7
Mogrunda obscura	62	31	0	0	62	7,170
Periophthalmus cantonensis	46	23	0	0	46	7
Tridentiger obscurus	44	22	0	0	44	7
Anabantidae						
Anabas testudineus	48	24	0	0	48	166
Betta splendens	42		6	0	36	171
” ”	42	21				172
Colisa fasciatus	48	24	0	0	48	110,166
Ctenopoma ansorgei		24				83
Macropodus opercularis		21				83,171
Trichogaster trichopterus		24				83
Mastacembellidae						
Rhynchobdella aculeata	48	24	0	0	48	166
Pleuronectiformes						
Bathidae						
Paralichthys olivaceus	46	23	0	0	46	7
Xystreurys liolepsis	48		0	0	48	36

TELEOSTEI (Continued)

CLASSIFICATION	CHROMOSOME NUMBER		CHROMOSOME TYPE			REFERENCES
	$2n$	n	M	SM	A	
Pleuronectidae						
Kareius bicoloratus	48	24	0	0	48	7
Limanda yokohamae	48	24	0	0	48	7
Pleuronectes platessa	48		0	0	48	173
Pleuronichthys verticalis	48		0	0	48	36,46
Tetraodontiformes						
Balistidae						
Stephanolepis cirrhifer	34	17	0	0	34	7
Rudarius ercodes	36	18	0	0	36	7
Novodon modestus	40	20	0	0	40	7

REFERENCES FOR CHECK-LIST OF FISH CHROMOSOMES

* 1. Van Der Strict: *Bull. Acad. Roy. Belg. Ser.,* 3: 1895.

* 2. Van Der Strict: *Bull. Acad. Roy. Belg. Ser.,* 30: 1896.

* 3. Van Der Stict: *Archives de Biologie, 14:* 1896.

* 4. Sobotta: *Archiv fur mikroskopische Anatomie,* 50: 1897.

* 5. Cerfontaine: *Acad. Roy. Belg.,* 1905.

* 6. Cerfontaine: *Archives de Biologie, 22:* 1905.

7. Nogusa, S.: A comparative study of the chromosomes in fishes with particular considerations on taxonomy and evolution. *Memoirs of the Hyogo Univ. of Agriculture,* 3 (1) : 1, 1960.

8. Howell, W.M., and Boschung, H.T.: Chromosomes of the lancelet, *Branchiostoma floridae* (Order Amphioxi). *Experientia, 27:* 1495. 1971.

9. Nogusa, S.: Study of chromosomes in two species of hagfishes. *Dubutsugaku Zasshi. Zoological Mag. 66:* 307, 1957.

10. Taylor, Kenneth, M.: The chromosomes of some lower chordates. *Chromosoma, 21:* 181, 1967.

* 11. Retzius; *Biol. Foren. Stockholm 2:* 1890.

* 12. Schreiner: *Anatomischer Anzeiger, 24:* 1904.

* 13. Schreiner: *Archives de Biologie, 21:* 1904.

* 14. Schreiner: *Archiv ful zellforschung, 1:* 1908.

15. Sasaki, Motomichi, and Hitotsumachi, Shinya: Notes on the chromosomes of a fresh water lamprey, *Entosphenus reissneri:* (Cyclostomata). *Chromosome Inf. Serv., 8:* 22, 1967.

16. Howell, W.M., and Duckett, C.R.: Somatic chromosomes of the lamprey, *Ichthyomyzon gagei* (Agnatha: Petromyzonidae) . *Experientia, 27:* 222, 1971.

17. Howell, W.M., and Denton, T.E.: Chromosomes of ammocoetes of the Ohio brook lamprey, *Lampetra aepyptera. Copeia,* No. 2: 393, 1969.

18. Potter, I.C.; Robinson, E.S., and Walton, Shirley, M.: The mitotic chromosomes of the lamprey *Mordacia mordax* (Agnatha: Petromyzonidae). *Experientia, 24:* 966, 1968.

19. Robinson, E.S., and Potter, I.C.: Meiotic chromosomes of *Mordacia praecox* and a discussion of chromosome numbers in lampreys. *Copeia,* No. 4: 824, 1969.

20. Potter, I.C., and Rothwell, B.: The mitotic chromosomes of the lamprey, *Petromyzon marinus L. Experientia, 26:* 429, 1970.

* 21. Kastschenko: *Zeitschrift fur mikros-anatomische Forschung, 50:* 1890a.

*Taken from Makino, Sajiro: *An Atlas of the Chromosome Numbers in Animals,* 2nd ed. Iowa, Iowa State College Press, 1951.

**Taken from Gyldenholm, A.O., and Scheel, J.J.: Chromosome numbers of fishes. I. *J. Fish. Biol.,* 3: 479, 1971.

* 22. Kastschenko: *Zeitschrift fur mikos-anatomische Forschung, 50:* 1890b.

* 23. Moore: *Anatomischer Anzeiger, 9:* 1894.

* 24. Moore: *Quarterly Journal of Microscopical Science, 38:* 1895.

* 25. Farmer and Moore: *Quarterly Journal of Microscopical Science, 48:* 1904.

* 26. Rawitz: *Archiv fur mikroskopische Anatomie, 53:* 1899.

* 27. Cerruti: *Att. Acad. Sci.* 13: 1908.

* 28. Matthey: *Comptes Rendus des Seances de la Societe de Biologie* (Paris), *126:* 1937.

* 29. Matthey: *Sci. Genetica,* 3: 1947.

* 30. Schreiner: *Archives de Biologie, 22:* 1907.

31. Nygren, Axel; Nilson, Birgit, and Jahnke, Margarette: Cytological studies in Hypotremata and Pleurotemata (Pisces). *Hereditals 67:* 275, 1971.

* 32. Makino: *Cyt. Fujii Jub.* 1937.

* 33. Kastschenko: *Zeitschrift fur wissenschaftliche Zoologie, 50:* 1890.

34. Ohno, S.; Muramoto, J.; Stenius, C.; Christian, L., and Kittrell, W. A.: Microchromosomes in Holocephalian, Chondrostean and Holostean fishes. *Chromosoma, 26:* 35, 1969.

* 35. Wickbom: *Hereditas,* 31: 1945.

36. Ohno, S., and Atkin, N.B.: Comparative DNA values and chromosome complements of eight species of fishes. *Chromosoma,* 18: 455, 1966.

* 37. Murry: *Anatomischer Anzeiger, 29:* 1960.

* 38. Agar: *Quarterly Journal of Microscopical Science, 57:* 1911.

* 39. Agar: *Quarterly Journal of Microscopial Science, 58:* 1912.

40. Denton, T.E., and Howell, W.M.: Chromosomes of the African polypterid fishes, *Polypterus palmas and Calamoichthys calabaricus. Experientia,* 1972.

41. Chiarelli, B.; Ferrantelli, O., and Cucchi, C.: The caryotype of some teleostean fish obtained by tissue culture in vitro. *Experientia, 25:* 426, 1969.

42. Sick, K.: Haemoglobin patterns and chromosome numbers of American, European, and Japanese eels (Anguilla). *Nature,* 193: 1001, 1962.

* 43. Rodolico: *Publ. Staz. Zool. 8:* 1933.

44. Mayers, Lyndon J., and Roberts, Franklin L.: Chromosomal homogeneity of five populations of alewives, *Alosa pseudoharengus. Copeia,* No. 2.: 313, 1969.

45. Roberts, Franklin L.: Cell structure of fibroblasts from *Clupea harengus gonads. Nature, 212:* 1592, 1966.

---

*Taken from Makino, Sajiro: *An Atlas of the Chromosome Numbers in Animals,* 2nd ed. Iowa, Iowa State College Press, 1951.

**Taken from Gyldenholm, A.O., and Scheel, J.J.: Chromosome numbers of fishes. I. *J. Fish. Biol.,* 3: 479, 1971.

46. Ohno, S.; Wolf, U., and Atkin, Niels: Evolution from fish to mammals by gene duplication. *Hereditus, 59:* 169, 1967.
47. Nayyar, R. P.: Karyotype studies in the genus *Notopterus. Genetica, 36:* 398, 1965.
48. Srivastava, M.D.: The structure and behavior of chromosomes in six fresh water teleosts. *Cellule, 65:* 93, 1964.
49. Simon, R.C.: Chromosome morphology and species evolution in the five North American species of Pacific salmon (Oncorphynchus). *J. Morphol., 112*(1):77, 1963.
* 50. Makino: *Zoological Magazine (Japan), 49:* 1937.
51. Sasaki, Motomichi; Shinya, Hitotsumachi; Makino, Sajiro, and Terao, toschiro: A comparative study of the chromosomes in the chum salmon, the Kokanee salmon and their hygrids. *Caryologia, 21* (4): 389, 1968.
52. Svardson, G.: Chromosome studies on Salmonidae. *Rept. Inst. Freshwater Res.* Drottingholm, *23:* 1, 1945.
53. Nygren, Axel; Nillson, Birgit, and Jahnke, Margarete: Cytological studies in *Salmo trutta* and *Salmo alpinus. Hereditas,* 67: 259, 1971.
* 54. Pomini: *Cci. Genet. 1:* 1939.
55. Simon, R.C., and Dollar, A.M.: Cytological aspects of speciation in two North American teleosts, *Salmo gairdneri* and *Salmo clarki lewisi. Canad. J. Genet. Cytol. 5:* 43, 1963.
* 56. Bohm: *Sitz. Ges. Morph. Phys. Minchen., 7:* 1891.
57. Heckman, J.R.; Allendorf, F.W., and Wright, J.E.: Trout leucocytes: growth in oxygenated cultures. *Science, 173:* 246, 1971.
58. Kobayasi, Hiromu; Kawashima, Yasuyo, and Takeuchi, Naomasa: Comparative chromosome studies in the genus *Carassius,* especially with a finding of polyploidy in the ginbuna *(C. auratus langsdorfii). Japanese Journal of Ichthyology, 17* (No. 4): 153, 1970.
59. Ohno, S.; Stenius, C.; Faist, E., and Zenzes, M.T.: Postzygotic chromosomal rearrangements in rainbow trout *(Salmo irideus* Gibbons). *Cytogenetics, 4:* 117, 1965.
60. Roberts, Franklin L.: Atlantic salmon *(Salmo salar)* chromosomes and speciation. *Trans. Amer. Fish. Soc.,* No. 1: 105, 1970.
* 61. Prokofieva: *Cytologia* (Tokyo): *5,* 1934.
62. Boothroyd, E.R.: Chromosome studies on three Canadian populations of Atlantic salmon, *Salmo sala* L. *Can. J. Genet. Cytol., 1:* 161, 1959.
63. Roberts, Franklin L.: Chromosomal polymorphism in North American landlocked *Salmo salar, Can. J. Genet. Cytol., 10:* 865, 1968.

---

*Taken from Makino, Sajiro: *An Atlas of the Chromosome Numbers in Animals,* 2nd ed. Iowa, Iowa State College Press, 1951.

64. Ross, H.: The chromosomes of *Salmo salar. Chromosoma, 21:* 472, 1967.
65. Nygren, Axel; Nillson, Birgit, and Jahnke, Hargarette: Cytological studies in Atlantic salmon *(Salmo salar). Acad. reg. scient Upsaliensis, 12:* 21, 1968.
* 66. Blanc: *Ber. Nat. Ges. Freiburg, 8:* 1894.
67. Wright, J.E.: Chromosome numbers in trout. *Progressive Fish Culturist, 17* (4): 172, 1955.
68. Wahl, R.W.: Chromosome morphology in lake trout *Salvelinus namaycush. Copeia,* (1) : 16, 1960.
69. Nygren, Axel; Nillson, Birgit, and Jahnke, Margarette: Cytological studies in *Thymallus thymallus* and *Coregonus albula. Hereditas, 67:* 269, 1971.
70. Booke, Henry E.: Cytotaxonomic studies of the Coregonine fishes of the Great Lakes, USA: DNA and karyotype analysis. *J. Fish. Res. Bd Canada, 25* (8): 1667, 1968.
71. Kupka, E.: Chromosomale Verschiedenheiten bei schweizerischen Coregonen. *Rev. Suisse Zool., 55:* 285, 1948.
72. Karbe, L.: Die chromosomen verhaltnisse bei den Coregonen des bodensees und einiger weiterer voralpiner seen, ein beitrag zum probelm der speziation in der gattung *Coregonus. Z. Zool. Syst. Evol., 2:* 18, 1964.
73. Nygren, Axel; Leijon, Urban; Nillson, Birgit, and Jahnke, Margarette: Cytological studies in *Coregonus* from Sweden. *Acad. reg. scient. Upsaliensis, 15:* 5, 1971.
74. Viktorovsky, R.M.: Chromosome sets of *Coregonus peled* and *C. lavaretus baunti. Tsitologiya, 6:* 636, 1964.
75. Bargetzi, J.P.: Application de methodes d'analyse biochimique a un probleme taxonomique: les Coregones du lac de Neuchatel. *Schweiz. Z. Hudrol., 22:* 641, 1960.
76. Nygren, Axel; Nilsson, Birgit, and Jahnke, Margarete: Cytological studies in the smelt (Osmerus eperlanus L.) *Hereditas, 67:* 283, 1971.
77. Chen, T.R.: Karyological heterogamety of deep sea fishes. *Postilla,* No. 130, 1, 1969.
78. Chen, T.R., and Ebeling, A.W.: Probable male heterogamety in the deep sea fish *Bathylagus wesethi. Chromosoma* (Berl.), *18:* 88, 1966.
79. Davisson, M.T.: Karyotypes of the teleost family Esocidae. *J. Fish. Res. Bd. Canada, 29:* 579, 1972.
80. Prakken, R.J.; Bekendam, R.J., and Peters, G.A.: The chromosomes of *Esox lucius* L., *Genetica, 27:* 484, 1955.

*Taken from Makino, Sajiro: *An Atlas of the Chromosome Numbers in Animals,* 2nd ed. Iowa, Iowa State College Press, 1951.

81. Nygren, Axel; Edmund, Per; Hirsch, Ulf, and Ahsgren, Lars: Cytological studies in perch (*Perca fluviatilis* L.), Pike (*Esox lucius* L.), Pike-perch (*Lucioperca luciopera* L.), and Ruff (*Acerina cernua L.). Hereditas, 59:* 518, 1968.

* 82. Foley: *Biological Bulletin, 50:* 1926.

** 83. Post, A.: Vergleichende Utersuchungen der Chromosomenzahlen bei Susswasser-Teleosteern. *Z. Zool. Syst. Evol., 3:* 47, 1965.

84. Muramoto, Jun-ichi; Ohno, Susumu, and Atkin, Niels B.: On the diploid state of the fish order Ostariophysi. *Chromosoma, 24:* 59, 1968.

85. Scheel, J.J., and Christensen, B.: The chromosomes of the two common neon tetras. *Trop. Fish Hobbyist, 19:* 24, 1970.

86. Scheel, J.J.: The chromosomes of the third neon tetra. *Trop. Fish Hobbyist,* July, 1972.

** 87. Lueken, W., and Foerster, W.: Chromosomen untersuchungen bei fischen mit einer vereinfachten Zelikulturtechnik. *Zool. Anx., 183:* 169, 1969.

88. Howell, W.M.: Somatic chromosomes of the black ghost knifefish, *Apteronotus albifrons* (Pisces, Apteronotidae). *Copeia,* No. 1: 191, 1972.

89. Nogusa, S.: Notes on the chromosomes of Abbottina rivularis (Cyprinidae). *Jap. Jour. Genet.,* 30: 10, 1954.

** 90. Lieder, U.: Chromosomen studien and Knochenfischen. *Z. Fischerei Hilfswiss.,* 3: 479, 1954.

91. Makino, S.: The chromosomes of the carp. *Cyprinus carpio,* including those of some related species of Cyprinidae for comparison. *Cytologia, 9:* 430, 1939.

92. Nayar, R.P.: Karyotype studies in seven species of Cyprinidae. *Genetica,* 35: 95, 1964.

93. Ohno, Susumu; Muramoto, Junichi, and Christian, Lawrence: Diploid-tetraploid relationship among old world members of the fish family Cyprinidae. *Chromosoma, 23:* 1, 1967.

94. Campos, Hugo H.: Karyology of three galaxiid fishes *Galaxias maculatus, G. Platei,* and *Brachygalaxias bullocki. Copeia,* No. 2, 1972.

* 95. Makino: *Jap. J. G.,* 9: 1934.

* 96. Makino: *Zoological Magazine* (Japan), *52:* 1940.

* 97. Makino: *Cytologia, 12:* 1941.

98. Kobayasi, Hiromu: A chromosome study in inter-family hybrids between the Funa and the loach. *The Nucleus, 8:* 1, 1965.

*Taken from Makino, Sajiro: *An Atlas of the Chromosome Numbers in Animals,* 2nd ed. Iowa, Iowa State College Press, 1951.

**Taken from Gyldenholm, A.O., and Scheel, J.J.: Chromosome numbers of fishes. I. *J. Fish. Biol.,* 3: 479, 1971.

99. Legendre, Pierre, and Steven, David M.: Denombrement des chromosomes chez quelques cyprins. *Naturaliste Can., 96:* 913, 1969.
100. Ojima, Y. and Hitotsumachi, S.: Cytogenetic studies in lower vertebrates. *Japan. J. Genetics, 42* (3): 163, 1967.
101. Ohno, S.; Wolf, U., and Atkin, Niels: Evolution from fish to mammals by gene duplication. *Hereditas, 59:* 169, 1967.
102. Nayyar, R.P.: Karyotype studies in two cyprinids: *Cytologia, 27:* 229, 1962.
103. Srivastava, M.D., and Kaur, D.: The structure and behavior of chromosomes in six fresh water teleosts. *Cellule, 65:* 93, 1964.
104. Lieppman, Michael, and Hobbs, Clark: A karyological analysis of two cyprinid fishes, *Notemigonus crysoleucas* and *Notropis lutrensis. Texas Reports on Biology and Medicine., 27* (2): 427, 1969.
105. Denton, T.E., and Howell, William M.: A technique for obtaining chromosomes from the scale epithelium of teleost fishes. *Copeia,* No. 2: 392, 1969.
*106. Nogusa: *Jap. J. Genet., 19:* 1943.
107. Uyeno, Teruya, and Smith, G.R.: Tetraploid origin of the karyotype of catostomid fishes. *Science, 175:* 144, 1972.
108. Hitosumachi, Shinya; Sasaki, Motomichi, and Ojima, Yoshio.: A comparative karyotype study in several species of Japanese loaches (Pisces, Cobitidae) . *Japan. J. Genetics,* 44 (3) : 157, 1969.
109. Srivastava, M.D.L., and Das, Bhagwan: Somatic chromosomes of teleostean fish. *J. Hered.,* 57, 1969.
110. Nayyar, R.P.: Karyotype studies in thirteen species of fishes. *Genetica, 37:* 78, 1966.
**111. Karbe, L.: Cytologische Untersuchungen der Sterilitatsers scheinungen bei anatolischen Zahnkarpfen, ein Betrag zum Speziations problen. *Mitt. Hamb. Zool. Mus. Inst., 59:* 73, 1961.
**112. Scheel, J.J.: Taxonomic studies of African and Asian toothcarps (Rivulinae). *Vidensk. Medd. Dansk naturhist. For., 129:* 123, 1966.
**113. Scheel, J.J.: Rivulins of the Old World. *T.F.H. Publ.* U.S.A., 1968.
**114. Scheel, J.J.: *Aphyosemion franzwerneri* and *Aphyosemion celiae,* two new rivulins from Cameroon. *Trop. Fish Hobbyist, 19:* 48, 1971.
**115. Scheel, J.J.: A new species of *Aphyosemion* (Cyprinodontidae) from Fernando-Poo. *Rev. Zool. Bot Afr., 78:* 332, 1968.
**116. Scheel, J.J.: Notes on the taxonomy of *Austrofundulus dolichopterus* and other annual rivuline species of the new worls. *J. Am. Killifish Assoc., 6:* 8, 1969.

---

*Taken from Makino, Sajiro: *An Atlas of the Chromosome Numbers in Animals,* 2nd ed. Iowa, Iowa State College Press, 1951.

**Taken from Gyldenholm, A.O., and Scheel, J.J.: Chromosome numbers of fishes. I. *J. Fish. Biol.,* 3: 479, 1971.

**117. Scheel, J.J.: *In Susswasser fische aus aller welt.* Sterba, G., II. 358, 1970.
118. Chen, T.R.: A comparative chromosome study of twenty killifish species of the genus *Fundulus* (Teleosteii Cyprinodontidae). *Chromosoma, 32:* 436, 1971.
119. Chen, T.R., and Ruddle, F.H.: A chromosome study of four species and a hybrid of the killifish genus (Cyprinodontidae). *Chromosoma, 29:* 255, 1970.
120. Chen, T.R., and Ebeling, A.W.: Heterogamety in teleostean fishes. *Trans. Amer. Fish. Soc.,* No. 1: 131, 1970.
*121. Moenkhaus: *Am. J. Anat.,* 3: 1904.
*122. Pinney: *Journal of Morphology, 31:* 1918.
123. Chen, T.R.: Fish chromosome preparation: air-dried displays of cultured ovarian cells in two killifishes *(Fundulus). J. Fish. Res. Bd. Can.,* 27 (No. 1): 158, 1970.
124. Setzer, Paulette Y.: An analysis of a natural hybrid swarm by means of chromosome morphology. *Trans. Amer. Fish. Soc.,* No. 1: 139, 1970.
*125. Geiser: *Biological Bulletin, 47:* 1924.
*126. Wickbom: *Hereditas, 29:* 1943.
*127. Winge: *Journal of Genetics, 12:* 1922.
*128. Winge: *Journal of Genetics, 13:* 1923.
*129. Vaupel: *Journal of Morphology,* 47, 1929.
*130. Iriki: *Zoological Magazine, 44,* 1932.
*131. Iriki: *Proceedings of the Imperial Academy* (Toyko), *8:* 1932.
*132. Meyer: *Journal of Genetics, 36:* 1938.
**133. Leuken, W.: Chromosomenzahlen bei *Orestias* (Pisces, Cyprinodontidae). *Zool. Mus. Inst., 60:* 195, 1962.
*134. Goodrich: *Journal of Experimental Zoology,* 49: 1927.
*135. Iriki: *Science Reports of the Tokyo Bunrika Deigaku,* 1: 1932.
136. Sriramulu, V.: A note on the chromosomes of *Oryzias melastigma. Current Science,* 28: 117, 1959.
137. Subrahmanyam, K., and Ramamoorthi, K.: A karyotype study in the estuarine worm eel, *Moringua linearis* (Gray). *Science and Culture, 37:* 201, 1971.
138. Chen, T.R., and Ebeling, A.W.: Karyological evidence of female heterogamety in the mosquitofish, *Gambusia affinis. Copeia,* No. *1:* 70, 1968.
139. Roberts, Franklin L.: A chromosome study of *Gambusia affinis holbrookii. Copeia,* No. 2: 239, 1965.

---

*Taken from Makino, Sajiro: *An Atlas of the Chromosome Numbers in Animals,* 2nd ed. Iowa, Iowa State College Press, 1951.

**Taken from Gyldenholm, A.O., and Scheel, J.J.: Chromosome numbers of fishes. I. *J. Fish. Biol.,* 3: 479, 1971.

140. Campos, Hugo H., and Hubbs, Clark: Cytomorphology of six species of gambusiine fishes. *Copeia,* No. 3: 566, 1971.
141. Drewry, George E.: App. I. Chromosome number. *Bull. Texas Memorial Museum.,* No. 8: 67, 1964.
142. Wickbom, T.: Cytological studies on the family Cyprinodontidae. *Hereditas, 29:* 1, 1943.
*143. Ralston: *Science, 78:* 1934.
*144. Ralston: *J. Morphol., 56:* 1934.
*145. Friedman and Gordon: *American Naturalist, 68:* 1934.
146. Prehn, Lilliam M., and Rasch, Ellen M.: Cytogenetic studies of *Poecilia* (Pisces) I. Chromosome numbers of naturally occurring Poeciliid species and their hybrids from Mexico. *Can. J. Genet. Cytol., 11:* 880, 1969.
**147. Winge, O.: A peculiar mode of inheritance and its cytological explantation. *J. Genet., 12:* 137, 1922.
148. Schultz, Jack R.: Gynogenesis and triploidy in the viviparous fish *Poeciliopsis. Science, 157:* 1564, 1967.
149. Schultz, Jack R.: Hybridization, unisexuality and polyploidy in the teleost *Poeciliopsis* (Poeciliidae) and other vertebrates. The *American Naturalist, 103* (934): 605, 1969.
**150. Freye, H.A.: Das chromosomebild von *Xiphophorus helleri, Platypoecilius maculatus* und den bastarden. (Xiphophorini, Pisces). *Biol. Zbl. Dtsch., 86:* 267, 1967.
*151. Moenkhaus: *Am. J. Anat., 3:* 1904.
152. Chen, T.R., and Reisman, H.M.: A comparative chromosome study of the North American species of sticklebacks (Teleostei: Gasterosteidae). *Cytogenetics, 9:* 321, 1970.
153. Muramoto, Jun-Ichi, and Igavashi, Kiyoshi: A preliminary note on the chromosomes and enzymatic patterns of three forms of sticklebacks. *Journal of the Faculty of Science* (Hokkaido Univ. Series VI, Zoology), *17* (1): 266, 1969.
154. Muramoto, Jun-chi; Igarashi, Kiyoshi; Ithh, Masahir, and Makino, Sajiro: A study of the chromosomes and enzymatic patterns of sticklebacks of Japan. *Proc. Japan Acad., 45:* 803, 1969.
*155. Makino: *Jap. Jour. Geentics, 9:* 1934.
156. Nogusa: *Jap. Jour. Genetics, 25:* 1950.
*157. Makino: *Zoological Magazine* (Japan), *49:* 1937.
*158. Hann: *Journal of Morphology, 43:* 1927.
159. Starmach, Janusz: Chromosomes of *Cottus peocilopus* and *Cottus gobio. Acta Hydrobiologica, 9* (3/4): 301, 1970.

---

*Taken from Makino, Sajiro: *An Atlas of the Chromosome Numbers in Animals,* 2nd ed. Iowa, Iowa State College Press, 1951.

**Taken from Gyldenholm, A.O., and Scheel, J.J.: Chromosome numbers of fishes. I. *J. Fish. Biol.,* 3: 479, 1971.

160. Nogusa, Shyunsaku: Chromosome studies in Pisces, VI. The X-Y chromosomes found in *Cottus pollux* Gunther (Cottidae). *Hokkaido Univ. Journal of the Faculty of Science* (Zoology), *13:* 289, 1957.
161. Subrahmanyam, K., and Natarajan, R.: A study of the somatic chromosomes of *Therapon cuvier* (Teleostei: Perciformes). *Proceedings of the Indian Academy of Sciences.* LXXII: 288, 1970.
162. Roberts, Franklin L.: A chromosome study of twenty species of Centrarchidae. *J. Morphol.,* 115 (3): 410, 1964.
163. Becak, W.: Becak, M.L., and Ohno, S.: Intraindividual chromosomal polymorphism in green sunfish as evidence of somatic segregation. *Cytogenetics,* 5: 313, 1966.
164. Bright, W.M.: Spermatogenesis in sunfish. *Transactions, Kentucky Academy of Science, 8:* 37, 1940.
165. Regan, James D.; Sigel, Michael M.; Lee, William H.; Llamas, Kirsten A., and Beasley, Annie R.: Chromosomal alterations in mraine fish cells in vitro. *Can. J. Genet. Cytol., 10:* 448, 1968.
166. Kaur, D., and Srivastava, M.D.L.: The structure and behavior of chromosomes in five fresh water teleosts. *Carynologia, 18* (2): 181, 1965.
167. Natarajan, R., and Subrahmanyam, K.: A preliminary study on the chromosomes of *Tilapia mossambica* (Peters). Curr. Sci., *37* (9): 262, 1968.
168. McPhail, J.D., and Jones, R.L.: A simple technique for obtaining chromosomes from teleost fishes. *J. Fish. Res. Bd. Canada, 23* (5): 767, 1966.
169. Chen, T.R.: Chromosomes of the goby fishes in the genus *Gillichthys. Copeia,* No. 1: 171, 1971.
170. Nogusa, Shunsaku: Chromosome studies in Pisces IV. The chromosomes of *Mogrunda obscura* (Gobiidae), with evidence of male heterogamety. *Cytologia,* 20: 11, 1955.
171. Svardson, G., and Wickbom, T.: The chromosomes of two species of Anabantidae (Teleostei) with a new case of sex reversal. *Hereditas, 28:* 212, 1942.
*172. Bennington: *Journal of Morphology, 60:* 1936.
173. Barker, C.J.: A method for the display of chromosomes of Plaice, *Pleuronectes platessa,* and other marine fishes. *Copeia,* No. 2: 365, 1972.

*Taken from Makino, Sajiro: *An Atlas of the Chromosome Numbers in Animals,* 2nd ed. Iowa, Iowa State College Press, 1951.

CHAPTER 6

EVOLUTION OF THE FISH KARYOTYPE

IT HAS ALREADY BEEN STIPULATED that the fish karyotype is the basic chromosome set of a species in which the chromosomes are characterized as to form, size and number. In addition to pathological studies and determining the effects of chemicals on chromosomes, the karyotype is a useful tool in the hands of the systematist who desires something more than morphological criteria for defining species relationships. In the evolutionary sense, studies would be incomplete without cytotaxonomic data to reinforce supposed relationships within natural groups based on morphological criteria. The fish karyotype, like any morphological character, is subject to some variability. With modern methods, however, this variability can be held to an absolute minimum. The cytotaxonomy of fishes is still in its infancy. Recent works have begun to open up this avenue of study and the future holds a great deal of promise.

Changes in Chromosome Number

Chromosome numbers in fishes range from $2n = 16$ *to* $2n = 174$ with the majority of fishes having numbers between forty and sixty. There is no evidence that basic numbers are connected to phylogenetic positions of major groups but within the narrower limits of family and genus there does seem to be a tendency for a reduction in chromosome number to parallel specialization. Also, there is a direct correlation between change in chromosome number and change in chromosome morphology. It has been surmised by some cytogeneticists, for example, that primitive or less specialized fishes have higher chromosome numbers and more acrocentric chromosomes while their more advanced relatives have more metacentrics. Mayers and Roberts (1969) implied that the alewife is a relatively primitive member of the order Salmoniformes because they contain forty-eight acrocentric chromosomes as compared to other members of the order with numerous metracentrics. From the listing in the preceeding chapter it is shown that *Fundulus parvipinnis* has forty-eight acrocentric chromosomes whereas the more advanced *Fundulus chrysotus* has thirty-four chromosomes, fourteen of which are metacentric. Numerous such examples can be shown within other families where karyotypes for several species have been done. This relationship, however, does not always hold true. Before a general trend can be established, many karyotypes will have to be made within the framework of other families and genera.

Both aneuploidy and euploidy are exemplified in fishes but occurrences are rare. Aneuploid organisms have chromosome numbers other than the exact multiple of the gametic number. In such cases, gametes would arise with deficiences and duplications which would render it inviable. Aneuploidy, therefore, reflects instability and would not be expected to occur to any significant extent among fishes. It will be remembered from the discussion of multiple sex chromosomes (Uyeno and Miller, 1971) that an undescribed killifish had somatic numbers of forty-eight for females and forty-seven for males. This condition arose through centric fusion of two acrocentric chromosomes, and while the sexes differ with respect to number, there does not seem to be a loss of functional genetic material in the male. This is a rare occurrence. Other fishes could very well become aneuploid due to non-disjunctions but their existence would expectedly be short lived.

Euploidy is the chromosome number of an organism that is an exact multiple of the gametic number. As with all eukaryotic organisms, diploid numbers are the rule in fishes. Monoploids, pentaploids and above are not known to occur in fishes. Tetraploids are thought to occur in some and these will be treated later in connection with discussions on DNA. Only one case of triploidy with cytological verification has been reported in fishes (Schultz, 1967). In this study, there were three all female monosexual strains of *Poeciliopsis* which occurs in the Rio Fuerte of

Sinaloa, Mexico. These strains were designated *Cx, Cy,* and *Cz.* The bisexual species, *Poeciliopsis lucida,* provides sperm for each of these forms. Form *Cy* is triploid and produces all female triploid offspring when mated to males of *Poeciliopsis latidens,* a test organism. These triploids are indistinguishable from the mother. Both *Cx* and *Cz* are diploid and express characteristics of both parents when mated to *P. lucida.* Both *P. latidens* and *P. lucida* are diploid species with forty-eight chromosomes. Triploid offspring thus have seventy-two chromosomes. It is suggested that the origin of the *Cy* strain resulted from a suppression of a cleavage stage and the production of diploid eggs which was then fertilized with sperm from *P. lucida.* Once triploid, meiosis was proceeded by endomitosis which is essentially a modified mitotic division that produces one nucleus having a multiple of the original chromosome number. The population is thought to be sustained by gynogenesis from triploid eggs. Cytological evidence to substantiate that endomitosis does occur in *Poeciliopsis* was given by Cimino (1972). Prior to meiosis in triploid, gynogenetic, all female forms, the triploid nucleus becomes hexaploid endomitotically. Recombination does not occur and the eggs receive a triploid complement identical to that of the mother. Sperm from a sympatric species then stimulates the ovum to develop without fusing with the nucleus.

Change in Chromosome Structure

The evolution of the karyotype and the creation of new species are due mostly to chromosome rearrangements. The repackaging of the DNA that is already on the chromosome can lead to effective reproductive barriers which control the type of offsping that can be produced. These chromosome rearrangements occur during interphase–prophase of meiosis and are detectable during late prophase and metaphase I. If the chromosome suffices one break, there is loss of DNA material in the form of a terminal deletion. Gametes which contain this broken chromosome are usually inviable because of gene deficiency. Should two breaks occur, a number of chromosome aberrations are possible. A break in two places on the same chromosome can lead to an intercalary deletion in which an internal segment is lost. If the internal segment becomes inverted so that there is no net loss of genetic material but simply a switching of ends, then the chromosome is said to contain an inversion. The inverted chromosome then forms a loop in order to pair with its homologue. Should crossing over occur during synapsis the resulting gametes containing the cross-over chromatids are inviable due to deficiences or duplications of genetic material. For this reason, inversions are aptly referred to as cross-over suppressors. Inversions have unquestionably played a major role in the evolution of new species by preventing mixings of gene pools between offspring with inversions and parental forms. Inversions can also produce visible changes in chromosome types. An inversion in an acrocentric chromosome in which the centromere is involved (pericentric inversion) can result in a metacentric, whereas an inversion in a metacentric involving a centromere will result in a large acrocentric.

If there are two breaks in which one is in a chromosome and the other in its homologue, and they subsequently undergo exchanges, then it is possible that one chromosome will end up with a deletion deficiency and the other with a duplication. Such a duplication of genetic material also leads to gametic inviability. However, if the duplicated portion represents redundant material from another chromosome, then it is possible for the rearranged complement to be maintained.

The most notable chromosome aberration produced in fishes is the translocation. In this type of aberration there are two breaks and subsequent exchanges between two non-homologous chromosomes. Such changes are termed reciprocal translocations and can be detected during synapsis as cross configuration quadrivalents. Even though 50 percent of the gametes are inviable, translocations are not as effective as inversions in preventing mixings of gene pools and giving rise to new species. The most common translocation in fishes is the Robertsonian type in which the centromeres of two non-homologous chromosomes fuse together producing a metacentric chromosome. Synaptic configurations typically reflect this occurrence as a long stranded trivalent containing three distinct segments in which the two end segments segregate to one pole and the middle segment to the other. Thus, metacentric chromosomes can be formed from acrocentric chromosomes. In like manner if the centromere of a metacentric is weak, then the chromosome can dissociate producing two smaller acrocentrics.

When Robertsonian fusions and inversions are considered together, numerous combinations of rearrangements are possible. An organism with one hundred acrocentric chromosomes, for example, may undergo fifty Robertsonian fusions to produce fifty metacentric chromosomes. These may undergo fifty inversions to produce fifty large acrocentrics which may subsequently undergo twenty-five fusions to produce twenty-five large metacentric chromosomes. With this type of mechanism it would seem that all one needed to do to detect these transformations would be to measure the arm lengths of all the chromosomes in karyotypes that were thought to be involved. Unfortunately, this would not be reliable since chromatin may be added or deleted at various times during the transformations. Centric fusions are more common than centric fragmentations or dissociations. It may well be that centric fusions are dominant in the evolution of specialized forms while dissociations occur in the reverse direction. In addition, translocations and inversions need not be thought of as involving all the chromosomes of a complement at the same time. Only parts of a complement, maybe only one chromosome, will have been altered at the precise time the organism is karyotyped.

Chromosomes and DNA Values

It is commonly held that all cells of a given species are characterized by a certain amount of DNA. If this is indeed the case, then to know the DNA value per cell for organisms of a group would contribute greatly to the understanding of phylogenetic relationships. Regretably, it isn't known whether or not the DNA that is measured is active or inert. Evolution, through natural selection, depends on mutation. Changes in gene material may occur as the result of change in chromosome number or structure, or by a change in the gene itself (point mutation). In the former two mutation types, the specific action of the structural gene (the cistron) might not be affected. In the latter type, gene function is changed and possibly those linked to it. In cases where the gene is duplicated, redundant DNA is produced. This redundant material may be lost, perform the same function as the material from which it came, or it may assume a different function and contribute to a new locus. Total DNA measurements do not differentiate these types but they do relate differences and similarities between some species. What these values show byond establishing trends for amounts of DNA between certain groups must await further refinements in methods of analysis.

Most of the DNA analysis work in relationship to chromosomes of fishes has been done by Ohno and co-workers (1966, 1967, 1968, and 1969). These workers interpret differences in DNA values of fishes to represent changes due to gene duplication. These changes may or may not be accompanied by changes in chromosome types and numbers. Ohno's premise contends that the most primitive fish karyotype consisted of forty-eight acrocentric chromosomes with a DNA value approximately 20 percent that of mammals. Mechanisms for changes due to gene duplication are envisioned to be of four types: 1) unequal exchange between sister chromatids during mitosis, 2) unequal crossing-over between two homologues during meiosis, 3) regional redundant duplication of DNA molecules in specific chromosomes and 4) by polyploidization, or tetraploidization, in which whole chromosomes are duplicated. The first three methods of gene duplication necessitate that duplicated gene loci be arranged in tandem on the same chromosome. In a disruptive fashion, the duplicated gene segments can undergo further crossing-over or the altered gene dosage ratio may interfere with normal function. In tetraploidization, the duplicated genes are located on different chromosomes. Allelic mutations alone are insufficient to account for the numerous changes that took place during vertebrate evolution. From this point of view, the gene duplication theory appears attractive. If the duplicated structural gene (cistron) remains under the control of the diffusable operon, then the duplications would increase the genome size without introducing new functions. Should the cistrons escape the influence of the operon, redundant duplicates would emerge and in time acquire new functions. At various stages of cellular development, different genes would be preferentially activated to accomplish a particular function.

The first report of DNA values for fishes was in 1966 (Ohno and Atkin, 1966) when eight species were compared with values for mammals. This consisted of the lungfish *Lepidosiren paradoxa* and seven actinopterygian forms. The DNA value for the lungfish was 3,540 percent that of mammals, the rainbow trout 80 percent, the goldfish 52 percent, the green sunfish 31 percent, and the swordtail, hornyhead turbot, and fantail sole had a value that was 19-23 percent that of mammals. In addition, the chromosomes ranged from thirty-eight metacentrics in the lungfish to forty-eight acrocentrics in the last three forms with combinations of both in the intermediate species. The high DNA value of the lungfish, coupled with the small number of large metacentric chromosomes gave reason to suspect that the chromosomes were polytenic and the genome had been increased through polyploidization. Since sex chromosomes act as a barrier to polyploidization, it was assumed that this multiplication occurred millions of years ago before the formation of sex chromosomes. At the other end of the scale, the swordtail, turbot, and sole were regarded as retainers of the original vertebrate genome. It was further presumed that the green sunfish and discus fish were triploid, and the goldfish represented pentaploid lineage while the rainbow trout demonstrated octaploid lineage which approaches the decaploid lineage of placental mammals.

Following the line of reason that polyploidization had occured in ancestral fishes, Atkin and Ohno (1967) obtained DNA values for four primitive chordates. This included the sea squirt, *Ciona intestinalis,* the lancelet, *Amphioxus lanceolatus,* the brook lamprey, *Lampetra planeri,* and the hagfish, *Eptatretus stoutii.* The DNA values for these forms were 6 percent, 17 percent, 38 percent and 78 percent respectively of the value for human cells. The diploid chromosome numbers were twenty-eight very small chromosomes for the sea squirt and forty-eight for the hagfish (Taylor, 1967). The numbers for the lancelet and brook lamprey were not reported. From associated forms, it was reported that *Branchiostoma belcheri* had thirty-two chromosomes and the lamprey *Entosphenus reissneri* had ninety-four to ninety-six minute chromosomes. From this it was concluded that the genomes of ancient organisms which gave rise to vertebrates contained very little DNA and that gene duplications occurred by polyploidization and by increases in chomosomal segments. Because of the higher DNA values in cyclostomes, it was suggested that these duplications occurred before the development of jaws. In the case of the hagfish, the chromosome complement was attained through all the proposed methods of duplication except polyploidization. In these ways, the DNA can be built up without altering the chromosome structure.

Ohno, *et al.* (1969) later presented DNA values for four primitive fishes consisting of the holocephalian ratfish, *Hydrolagus colliei* ($2n = 58 \pm$), the chondrostean shovelnose sturgeon, *Scaphirhynchus platorhynchus* ($2n = 112 \pm$), and two holosteans, the spotted gar, *Lepisosteus productus* ($2n = 68 \pm$), and the bowfin, *Amia calva* ($2n = 46 \pm$). In addition, microchromosomes were found in all species except the bowfin. The DNA values were 43 percent, 50 percent, 41 percent, and 37 percent that of mammals respectively. The sturgeon and gar were likened to some reptilian and avian species in that both contain microchromosomes and had DNA values 41 percent-50 percent that of mammals.

It was contended that evolution from crossopterygian fishes to labyrinthodont amphibians was polyphyletic, or represented several lineages, and that the crossopterygian ancestor of avian species had already attained characteristic genome sizes endowed with microchromosomes through tetraploidization. The gar and bowfin were regarded as left over relices among bony fishes because of the primitive morphological characteristics shared with crossopterygian forms.

Other teleosts with high chromosome numbers were regarded as tetraploids. The carp, goldfish, trout and salmon have high chromosome numbers and DNA values. If the forty-eight acrocentric chromosomes of the primitive ancestor became tetraploid, the amount of DNA would increase. The various morphologies of these forms could vary through random inversions and centric fusions.

Ohno, *et al.* (1967 reported on the DNA content of fishes which included two barb species, *Barbus tetrazona, B. fasciatus,* the goldfish, *Carassius auratus* and the carp, *Cyprinus carpio.* The first two members had diploid chromosome numbers of fifty and fifty-two and DNA values 20-22 percent that of placental mammals. The latter two cyprinids had diploid numbers of approximately 104 with DNA values 50-52 percent that of placental mammals. It was concluded that diploid-tetraploid relationships existed within these respresentatives of the family Cyprinidae, the barbs being diploid and the goldfish and carp tetraploid. The meiotic figures from the goldfish and carp contained mostly bivalents rather than quadrivalents. It was suggested that these fishes had a common ancestor that was tetraploid and that enough time had elapsed for the original four homologous sets to diverge through chromosome rearrangements into different homologous pairs. In this manner diploidization from a tetraploid was accomplished. The two barb species were reported to have originated from diploid ancestors with similar chromosome types and DNA values. Such suspected tetraploidy is confirmed in some cases as in the meiotic figures of the rainbow trout which show several multivalents. Additional evidence to substantiate gene duplication in these tetraploid forms have been demonstrated through electrophoretic isoenzyme patterns. The enzyme lactose dehydrogenase (LDH) is thought to be produced in its most elemental form from two different gene loci. In diploid fishes such as the smelt and pacific herring, only three bands are observable from skeletal and heart muscle. In the rainbow trout, which is suspected of being tetraploid, a total of five isoenzyme bands were found suggesting that tetraploidization of two alleles of the same gene locus produced two separate gene loci.

In similar fashion it was suggested that some of the primitive teleosts with DNA values approaching mammals had undergone a second duplication by tetraploidization. It was further reasoned that the ancient crossoptergian fishes that served as ancestors to amphibians had increased their DNA through tetraploidization while still aquatic.

Other examples of using DNA values to substantiate polyploidy have been reported. Uyeno and Smith (1972) reported increased DNA values of 50 percent that of placental mammals for fourteen species of catostomids which had $2n$ numbers ranging from ninety-six to 102. It was suggested that these forms evolved by tetraploidy from a cyprinid-like ancestor with a $2n$ number of fifty.

All workers do not agree that salmonid representatives are tetraploid. Rees (1964) studied the chromosomes and DNA in the Atlantic salmon *(salmo salar,* $2n = 60$) and the brown trout *(Salmo trutta,* $2n = 80$) to determine whether or not the organisms were polyploid. The DNA values were similar and the chromosome types were different which led to the conclusion that polyloidy was unlikely since chromosome fragmentation and fusion could explain the differences in chromosome types and sizes. This conclusion was based on previous studies by Svardson (1945) who suggested a basic haploid number of ten for the salmonid group making the Atlantic salmon a hexaploid and the brown trout an octaploid based on recurring numbers of chromosome types in the complements of both species.

Simon (1963) reported diploid numbers and morphologies of five species of *Oncorhychus* and also criticized the polyploid theory of Svardson. Disagreement in salmonid karyotypes prevented precise comparisons to be made. There was favor, however, in selecting Robertsonian fusions as the operative mechanism for the occurrence of the observed karyotypes of *Oncorhynchus.*

Muramoto and Ohno (1968) reported chromosomes and DNA values for seven species of fishes representing five families of the order Cypriniformes and two families of the order Siluriformes. The family Cyprinidae was represented by one species, the black shark, *Labeo chrysophekadion* which had a $2n$ number of fifty chromosomes and a DNA value of 40 percent that of placental mammals. Two species of the family Cobitidae had similar DNA values 29 percent and 27 percent but the khulli loach, *Acanthophthalmus khulli* had fifty chromosomes while the clown loach, *Botia macracantha* had ninety-six chromosomes, seventy being acrocentrics. In this instance, the number of chromosomes is increased without a change in DNA content. In the family Characinidae, the characin, *Chalceus macrolepidotus* had fifty-four chromosomes and 29 percent DNA while the piranha, *Serrasalmus hollandi* had sixty-four chromosomes and 48 percent DNA. In the Siluriformes, the channel catfish, *Ictarulus punctatus* of the family Ictarulidae had fifty-six chromosomes and 30 percent DNA and the armoured catfish, *Hypostomus plecostomus* had fifty-four chromosomes and 51 percent DNA.

All members studied were considered to be diploid having increased their DNA content above the basic 20 percent by unequal sister chromatid exchange during mitosis, unequal crossing-over between homologues during meiosis, and by regional duplication of chromosomal segments. These increases were accompanied by chromosomal rearrangements. The greatest increase in DNA was in the generalized or primitive forms. The armoured catfish and the pyranha had DNA values as high as the carp and goldfish which was previously studied, but were not considered to represent tetraploid species since they had 2*n numbers* of fifty-four and sixty-four respectively.

The most extensive study on DNA values of fishes was reported by Hinegardner (1968). Using a fluorometric technique he examined the haploid amount of DNA in ninety-seven fishes. All of these were teleost comprising forty families. The DNA content ranged from 0.4 to 4.4 picograms (10^{-12}gms.) or about 14 percent to 125 percent that of mammals where one picogram of DNA is approximately equivalent to 28 percent of the DNA of mammals (Ohno, 1970). Taylor divided these teleosts into two groups. Members having DNA contents between 4.4 and 1.2 picograms were regarded as more primitive and occupying a lower phylogenetic poistion than species with 0.8 to 0.45 picograms which represented more specialized forms and a higher phylogentic position. This was especially evident at the higher and lower values. Those with intermediate values showed intermediate degrees of variation and spcialization. The major conclusion drawn by Taylor was that more advanced species have less DNA than primitive species. Fishes with low amounts of DNA appear to be near the ends of evolutionary lines while fishes with high amounts of DNA represent evolutionary blind alleys. For some fishes there was not enough DNA to experiment with in changing environments. This could have possibly resulted in their extinction or else imposed limitations on potential evolutionary experimentations and adaptations.

Goin and Goin (1968) have discussed the evolution of DNA in fishes, in terms of the amount of DNA per nucleus, within the framework of evolutionary theories based on morphological structures. It was suggested that an elevated level of DNA per nucleus occurred in the protochordate stock that gave rise to vertebrates. According to Rendel (1965), there are three levels of genetic change in evolution. Type 1 postulates that a particular genome has a basic number of genes to code for certain enzymes. These genes may exist in allelic states but do not change entirely to encode completely different functions. This type of genome is restricted in producing variability. Type 2 involves a genome in which certain genes are no longer needed by the organism. The genes may either mutate to different genes with different enzyme producing capabilities, be lost from the genome, or simply remain in the inactive state. The latter of the three conditions seems remote since studies have shown that loss of enzyme function accompanies a decline in the amount of DNA (Brown, 1962). It is presumed that either case is caused by some environmental change. Type 2 evolution could result in slow moderate changes, but would seem inadequate to explain the widely diverse karyotypes that are existant among fishes. Rendel's Type 3 evolution involves the addition of new genetic material. This can arise by polyploidy, polyteny, serial duplication of chromosomal segments, or by other methods.

With these theories in mind the authors compared the amounts of DNA per nucleus for different animals. They cited 0.3 picograms (pg) of DNA per nucleus for Tunicata and 5.0 pg for Agnatha and suggested that there was a massive increase in the amount of nuclear DNA before the beginning of pisces evolution. The possibility was also considered that tunicates are degenerate forms completely removed from the evolutionary line. Nevertheless, primitive fishes do have far more DNA/nucleus than do the invertebrates. The Chondrichthyes had values of 5.5-6.7, not a significant jump from the lampreys. Values for the Osteichthyes, however, were 0.9-6 pg for Actinopterygii, and 100 pg for Sarcopterygii. Within the teleosts, values were highest for the more primitive forms and lowest for the more advances species. This supported Taylor's (1967) earlier findings. The high values for the lungfishes of the Sarcopterygii indicate that there was another massive increase in DNA between the ancestral fish form and higher vertebrates.

Polyploidy was viewed as the mechanism responsible for the major increases in DNA at the two sites indicated. Amphibians of today, however, are diploid organisms and not polyploid. In this regard, Swanson (1957) points out that polyploidy can produce diploidy (diploidization) through aberrations which reduce the homology between the sets of chromosomes present so that two of the four original homologues are eliminated from the genome.

Enjoining these findings with Rendel's theories, it was suggested that his Type 3 form of evolution was responsible for initiating major increases in cellular DNA from the protochordates to the fishes and that Type 2 was subsequently responsible for its decrease among the various ray finned groups. Type 3 was also responsible for the transition between the fish ancestor and amphibians with accompanying processes of diploidization.

Another example of how karyological studies and DNA values have helped in the study of cytotaxonomy has been in the positioning of the polypterids (Denton and Howell, 1972). This group consists of a single family containing the reedfish, *Calamoichthys calabaricus* and about nine species of bichirs of the genus *Polypterus.* Most classification schemes place the polypterids in the subclass Actinopterygii, superorder Chondrostei, with the sturgeons and paddlefishes. When the chromosomes of the reedfish and a representative bichir, *Polypterus palmas* were studied, it was found that both had $2n$ numbers of thirty-six. All the chromosomes are biarmed with the reedfish having thirty metacentrics and six submetacentrics and the bicher twenty-four metacentrics and twelve submetacentrics. Both specimens have the same number of arms but the arms differ in size. The only chondrostean specimen to be karyotyped is the shovelnose sturgeon, *Scaphirhynchus platorhynchus* which has 112 chromosomes of which forty-eight are microchromosomes. The polypterids do not have microchromosomes and their chromosomes are about three times longer than those of the sturgeon. Their chromosomes are also very different from the holostean gar, *Lepisosteus productus* ($2n = 68$) and bowfin, *Amia calva* ($2n = 46$)). The gar, like the sturgeon, has microchromosomes and smaller sized macrochromosomes. The bowfin lacks microchromosomes but has a higher $2n$ number and the chromosomes are smaller.

The chromosomes of *Calamoichthys* and *Polypterus* have more in common with the dipneustan lungfishes than with the actinopterygians. The lungfishes have from thirty-two to thirty-eight metacentrics which are about three times longer than those of the polyterids. Bachman, *et al.* (1972) have recently reported DNA values for *Polypterus palmas* of 11.7 picograms as compared to 1.7 pg for teleosts and 2.3-3.5 for holosteans and chondrosteans. The only surviving crossopterygian is the coelacanth *(Latimeria)* which has a DNA value 6.5 pg higher than the majority of the Actinopterygii which places it in about the same category with the polypterids. Unfortunately, the coelacanth has not been karyotyped.

The lungfishes have very large cells with 100-284 pg of DNA in each cell. The authors acknowledged that these findings in themselves were insufficient to establish definite relationships but did favor grouping the polypterids with the Sarcopterygii than with the Actinopterygii. There are not enough surving representatives and relatives of polypterids to determine with any substance how the karyotypes of these groups have evolved but from the size and structure of the chromosomes, and from the reported DNA values, it seems logical to position them separately from the Actinopterygii and the Dipneusti. In terms of DNA content, the polypterids and the coelacanth represent increases whereas the DNA in ray-finned fishes have declined. This increase, however, is not of the order found in lungfishes.

Expectedly, chromosome complements that differ are easily recognized but one must not get the impression that karyotypic studies should be done only to show disimmilarities. Karyotypic analysis can also show close relationships. Davisson (1972) karyotyped representatives of all five surviving species in the teleost family Esocidae including pikes, pickerels, muskellunge, and two interspecific hybrids and found that all had complements containing fifty acrocentric chromosomes. It was concluded from the similarity and stability of the esocid karyotypes that chromosomal rearrangements were not prerequisite to speciation within this group. Although species hybridize readily, they remain separate because of physical, behavioral or ecological barriers. One such barrier, for example, would be size differences between the large muskellunge and the smaller pickerels. Artificial crosses would eliminate these barriers and the remaining possibilities of gametic and developmental obstacles would be nullified if the karyotypes of the hybrids were proven to be similar. This was found to be the case with the esocids. For species whose ranges overlapped in nature, the species would be separately maintained by barriers such as size incompatibility or perhaps differences in spawning time and place. It was also noted by Davisson that the DNA values of esocids were 32-38 percent that of mammals and this reflects a diploid complement close to the ancestral fish species.

In a like vein, Mayers and Roberts (1969) reported similar karyotypes of forty-eight acrocentric chromosomes for five distinct populations of alewives, *Alosa pseudoharengus,* indicating that this species is comparatively unspecialized. It was suggested that the various populations were no older than twelve thousand years which approximates the time of the Wisconsin glaciation recession. This amount of time could have been insufficient to allow for chromosome variability in this species thus resulting in an isomorphic and stable karyotype. Lagler, *et al.* (1962) states that species in the area studied by Mayers and Roberts are probably a million years old and the ten thousand to twelve thousand years since the Wisconsin glaciation is associated with differentiation at no higher than the subspecies level.

Chromosomal Rearrangements

It has been suggested that chromosome complements of fishes are in general somewhat flexible. If this is true, and if Robertsonian evolution occurs in widespread fashion, then it would sometimes be expected that individuals of a species would have different chromosome numbers and types from other individuals of the same species. Cytological verification of such a happening was reported by Nogusa (1955) in the Japanese cyprinid *Acheilognathus rhombea.* Three different chromosome numbers were found in testicular material consisting of $2n = 44$ (four metacentrics and forty acrocentrics), $2n = 46$ (two metacentrics and forty-four acrocentrics), and $2n = 48$ (all acrocentrics). A single individual could thus have two or even three karyotypes. These observations were attributed to multiple formations through centric fusions. It was proposed that different combinants arose through sperm carrying different numbers of metacentrics.

In a more extensive study, Ohno, *et al.* (1965) cited cases of post-zygotic chromosomal rearrangements in the rainbow trout, *Salmo irideus.* Chromosome numbers in these specimens ranged from fifty-nine to sixty-four. Cells from spleen and kidney most often contained fifty-nine chromosomes, while liver parenchymal cells usually contained sixty-one to sixty-two. Since each somatic cell type is represented in the early embryo, it was suggested that this trend in chromosome number was established during a brief time in embryonic development. Upon examination of meiotic figures from both the ovary and testis, it was found that each contained eighteen presumptive bivalents and six obvious multivalents for some of the spreads. It was concluded that a Robertsonian type of chromosomal polymorphism prevailed within each individual with certain chromosome arms undergoing fusion and dissociation early in embryonic life. This subsequently produced somatic cells with different chromosome number and types.

Another case of intraindividual chromosomal polymorphism was aptly demonstrated in the green sunfish *Lepomis cyanellus* (Becak *et al.,* 1966). This species was selected to better solidify theories concerning the mechanism proposed to explain chromosomal rearrangements in the rainbow trout. North Carolina specimens having 48 acrocentrics were bred to West Virginia specimens with a diploid number of forty-six, including a homologous pair of large metacentrics. Specimens from these mixed populations contained $2n = 48$ (all acrocentrics), $2n = 46$ (two metacentrics, homozygous for a translocation), and $2n = 47$ (one metacentric, heterozygous for a translocation). All three types had 48 arms each. In meiotic spreads from the testis twenty-four bivalents, twenty-three bivalents, and twenty-two bivalents plus one presumptive trivalent were found. These findings were attributed to somatic segregation. It is presumed that most zygotes are heterozygous at the start of their development for some translocations. Somatic segregation tends to restore the homozygous condition for either the translocated types or for complements that are completely free of translocations. It would thus be expected that intermediate types would occur in some tissues. If the zygotes contain only homologous pairs, polymorphism would not be detected. Thus in the green sunfish it was presumed that specimens were heterozygous for a single Robertsonian translocation which results in chromosomal polymorphism through somatic segregation.

Looking at chromosomal polymorphism in another system, some interesting comparisons can be made between fish chromosomes cultured *in vitro* and those that are normally studied *in vivo*. It is known that cultured mammalian cells will eventually undergo chromosomal changes producing heteroploid types. This usually results after, but can occur before, approximately fifty passages and can result in aneuploidy, polyploidy, or structural chromosome aberrations. Some cell lines are more stable than others and have been maintained through subculture for years. Clem, *et al.* (1961) established a cell line from fin tissue of the blue sriped grunt, *Haemulon sciurus* that has now been in culture for about twelve years. Regan, *et al.* (1968) reported on the status of the chromosome complement after the Grunt Fin (GF) cell line had been cultured for nearly eight years with more than 250 passages. The cultured cells at that time contained forty-four acrocentric and two metacentric chromosomes with one of the acrocentrics containing a secondary constriction. Chromosomes in primary fin tissue contained forty-eight acrocentric chromosomes with two of the chromosomes sometimes having secondary constrictions. The obvious conclusion was that the GF cell line became homozygous for a Robertsonian translocation producing forty-six chromosomes. In the event that the GF cells were already homozygous for the translocation, the authors reported that one would expect $2n$ numbers of forty-six, forty-seven, and forty-eight in the subcultured samples. This is a different interpretation from those of Becak, *et al.* (1966) who postulated that heterozygous Robertsonian translocation should eventually produce homozygous, heterozygous, and no translocations whereas if the translocations were homozygous from the start, polymorphism would not be detected. Becak, *et al.* (1966) had also proposed that two chromosome arms which had existed as independent acrocentrics may unite to become a single metacentric. This was stated in connection with early zygotes and was not considered to be probable. The works of Regan, *et al.* (1968) points out that centric fusions in somatic cells are probable but at a much slower rate than that reported in mixed populations of green sunfish. Two different cell systems are involved however. Translocation rates could be much accelerated in embryonic tissues such as those of the sunfish study. To compare chromosomes from a fibroblast system with those from cells of a developing zygote, one must decide whether growing fibroblasts are more aligned with embryonic tissues or with differentiated and mature somatic tissues.

The most documented occurrence of chromosomal polymorphism occurs in *Salmo salar*. European species are made up of seventy-two arms consisting of twelve metacentric-submetacentrics, and forty-eight acrocentrics (Prokofieva, 1934, Svardson, 1945). In contrast, Boothyroyd (1959) reported seventy-two arms from Quebec, New Brunswick and Nova Scotia populations having sixteen metacentric-submetacentrics, and forty acrocentrics. Rees (1967) reported seventy-four arms in specimens from Wales containing sixteen submetacentrics and forty-two acrocentrics. Roberts (1968) reported model numbers of fifty-seven for Maine populations with culls giving a mode of fifty-six. In a later study, specimens from the Machias River (Maine) showed a sharp mode of fifty-four chromosomes, while those from the Narraguagus River (Maine) had modes of fifty-five and specimens from the Miramichi River (New Brunswick) exhibited modal numbers of fifty-six (Roberts, 1970). In all populations a relatively constant arm number of seventy-two was found.

In each of the above cases Robertsonian mechanisms could possibly be operative. Roberts (1968, 1970) states that *Salmo salar* chromosomes appear to be in a highly unstable state and are prone to chromosomal polymorphism. Whether or not this species is tetraploid (or having stemmed from a tetraploid) with subsequent fusions, dissociations and inversions is still conjectural. The only thing that is readily apparent is that the species as a whole contains different somatic chromosome numbers and these differences are very evident between populations.

Studies by Pegington and Rees (1966) were done with *Salmo salar* ($2n = 58$) and *Salmo trutta* ($2 = 80$) to determine if chromosome size is genetically determined or if it arises structurally through centric fusion or fragmentation The chromosomes of *S. salar* are characteristically longer than those of *S. trutta.* The largest metacentric chromosome was measured from each of the two species and the same type chromosomes measured in their hybrid It was found that the size did not change in the hybrid but remained constant It was thus concluded that the chromosome size was not genetically determined, nor was it determined as a consequence of polyploidy, but resulted from structural changes such as fusion or fragmentation.

Concluding Remarks

From the previous chapters, it can be seen that much has been learned about the fish karyotype in recent years Even though definitive conclusions cannot be made at this time relevant to all aspects of fish chromosomes, there are trends and mechanisms emerging that complement more factual knowledge concerning their number, structure, and evolution. Some of the features of fish chromosomes that are currently known are summarized as follows:

1. Most fishes have $2n$ numbers between forty and sixty with a median of about fifty. The numbers range from a low of sixteen in some tropical forms to a high of 174 in the lamprey.
2. Of the forty-one orders of fishes, chromosomes numbers and/or types are known for representatives of twenty-five. More karyological information is known for the families Salmonidae, Coregoninae, Cyprinidae, Cobitidae, Cyprinodontidae, Poeciliidae, Gasterosteidae, Centrarchidae and Gobiidae, than for any others.
3. Chromosome numbers are known for approximately six hundred of the estimated twenty thousand species of fishes. Of this number, karyotypes are known for about three hundred.
4. The most fundamental karyotype of fishes consists of approximately 48 acrocentric chromosomes and a DNA content of one picogram per cell or about 20 percent that of placental mammals.
5. There is a tendency for evolving groups of fishes to have fewer

chromosomes through metacentric formations with a corresponding decrease in size and amount of DNA.

6. Comparisons between chromosomal numbers and types are more reliable within the framework of the family and genus.
7. Triploidy and aneuploidy (trisomy) are represented in fish karyotypes but to a rare extent. Tetraploidy is thought to occur more frequently but exists in partially diploidized states. Complete tetraploidy showing sets of four homologous pairs has not been cytologically demonstrated for fishes.
8. The majority of fish karyotypes are isomorphic but some are heteromorphic with possible sex chromosomes. Most fishes are bisexual and either the male or the female may be heterogametic.
9. Major changes in chromosome type and number most probably occur through tetraploidy and serial duplication of chromosome segments. Major changes are most evident between the protochordates and primitive fishes and between crossopterygian ancestors and tetrapods.
10. Minor chromosome changes occur through deletion and duplication of specific loci as a result of unequal somatic sister chromatid crossovers, or unequal meiotic crossovers between homologues. These occur at all stages of evolution.
11. Chromosomes are rearranged through centric fusions, dissociations, or inversions and occur at all stages of phylogeny.
12. Gene duplication leads to redundancy. The redundant gene may be lost, function as the parent gene, or come to function in a slightly different way and eventually lead to the establishment of a new locus.
13. Chromosomal polymorphism occurs in fishes and is made possible as a consequence of translocations at times of somatic segregation.

REFERENCES

Anderson, T.: *Trans. N.Y. Acad. Sci.,* 13:130, 1951.

Atkin, Niels, B., and Ohno, Susumu: DNA values of four primitive chordates. *Chromosoma,* 23:10, 1967.

Bachmann, Konrad; Goin, Olive B., and Goin, Coleman J.: The nuclear DNA of *Polypterus. Copeia,* No. 2:363, 1972.

Barker, C. J.: A method for the display of chromosomes of *Plaice, Pleuronectes platessa,* and other marine fishes. *Copeia,* No. 2:365, 1972.

Becak, W.; Becak, M. L., and Ohno, S.: Intraindividual chromosomal polymorphism as evidence of somatic segregation. *Cytogenetics,* 5:313, 1966.

Berg, L. S.: Classification of fishes both recent and fossil. *Trav. Inst. Zool. Acad. Sci. USSR, 5:*87-517; Reprint, 1947.

Blair, Frank W.; Blair, Albert P. Brodkorb, Pierce; Cagle, Fred R., and

Moore, George A.: *Vertebrates of the United States,* 2nd ed. N.Y. McGraw, 1968.

Breder, Charles M.: *Field Book of Marine Fishes of the Atlantic Coast.* N.Y., Putnam, 1948.

Brown, G.W., Jr.: Urea cycle and cellular deoxyribonucleic acid content. *Nature, 194*:1279, 1962.

Boothroyd, E.R.: Chromosome studies on three Canadian populations of Atlantic salmon, *Salmo salar* L. *Can. J. Genet. Cytol., 1*:161, 1959.

Caspersson, T.; Zech, L.; Johansson, C., and Medest, E.J.: Identification of human chromosomes by DNA-binding fluorescent agents. *Chromosoma,* 30:215, 1970.

Catton, W.T.: Blood cell formation in certain teleost fishes. *Blood, 6* (1): 39, 1951.

Chen, T.R.: Fish chromosome preparation: Air-dried displays of cultured ovarian cells in two killifishes *(Fundulus). Journal Fisheries Research Board of Canada, 27* (1): 158, 1970.

Chen, T.R., and Ebling, A.W.: Probable male heterogamety in the deep-sea fish *Bathylagus westhi* (Teleostei: Bathylagidae. *Chromosoma, 18:* 88, 1966.

Chen, T.R., and Ebeling, A.W.: Karyological evidence of female heterogamety in the mosquitofish, *Gambusia affinis. Copeia,* No. 1: 70, 1968.

Chen, T.R.: Karyological heterogamety of deep-sea fishes. *Postilla,* No. 130: 1, 1969.

Cimino, Michael C.: Meiosis in triploid all-female fish *(Poeciliopsis, Poeciliidae). Science, 175:* 1484, 1972.

Clem, William L.; Moewus, Lisolette, and Sigel, Michael M.: Studies with cells from marine fish in tissue culture. *Proc. Soc. Exp. Biol. Med., 108:* 762, 1961.

Conger, A, and Fairchild, L.: A quick freeze method for making smear slides permanent. *Stain Technol., 28:* 281, 1953.

Devisson, Muriel Trask: Karyotypes of the teleost family Esocidae. *Journal Fisheries Research Board of Canada, 29* (5): 579, 1972.

Denton, Thomas E., and Howell, William M.: A technique for obtaining chromosomes from the scale epithelium of teleost fishes. *Copeia, 2:* 392, 1969.

Denton, T.E., and Howell, W.M.: Chromosomes of the African polypterid fishes, *Polypterus palmas* and *Calamoichthys calabaricus* (Pisces: Brachiopterygii) . Experientia (in press) .

Drewry, G.E.: App. I. Chromosome number. *Bull. Texas Memorial Museum,* No. 8: 67, 1964.

Dupraw, E.J.: *Proc. Natl. Acad. Sci., 53:* 161, 1965.

Dupraw, E.J.: *Nature, 209:* 577, 1966.

Ebeling, A.W., and Chen, T.R.: Heterogamety in teleostean fishes. *Trans. Amer. Fish. Soc.,* No. 1: 131, 1970.

Eddy, Samuel: *How to Know the Freshwater Fishes* (U.S. only). 2nd ed., Iowa, Brown, 1969.

Emmens, C.W.: Principles of aquarium management. In *Exotic Tropical Fishes.* N.J., T.F.H. Publications, M-3.00, 1962.

Endo, Akira and Ingalls Theodore H.: Chromosomes of the zebra fish. *J. Hered., 59:* 382, 1968.

Goin, Olive B., and Goin, Coleman J.: DNA and the evolution of the vertebrates. *The American Midland Naturalist; 80* (2): 289, 1968.

Grassé, P.P.: *Traité de Zoologie.* Paris, Masson et Cie, Tome XIII, 3 Vols., 1958.

Greenwood, Humphry P.; Rosen, Donn E.; Weitzman, Stanley H., and Myers, George S.: Phyletic studies of teleostean fishes, with a provisional classification of living forms. *Bulletin of the American Museum of Natural History, 131* (Article 4): 341, 1966.

Gyldenholm, A.O., and Scheel, J.J.: Chromosome numbers of fishes. *J. Fish. Biol.,* 3: 479, 1971.

Harrington, R.W., Jr.: Oviparous hermaphroditic fish with internal self-fertilization. *Science,* 135: 1749, 1961.

Heckman, Robert J., and Brubaker, Paul E.: Chromosome preparation from fish blood leucocytes. *The Progressive Fish-Culturist,* 32 (4) : 206, 1970.

Heckman, J.R.; Allendorf, F.W., and Wright, J.E.: Trout leucocytes: growth in oxygenated cultures. *Science, 173:* 246, 1971.

Hinegardner, Ralph: Evolution of cellular DNA content in teleost fishes. *The American Naturalist, 102* (928): 517, 1968.

Labat, Rene; Larrouy, Georges, and Malaspina, Liliane: Technique de culture des leucocytes de *Cyprinus carpio* L. *C. R. Acad. Sc. Paris, 264:* 2473, 1967.

Lagler, Karl F.; Bardach, John E., and Miller, Robert R.: *Ichthyology.* N.Y., Wiley, 1962.

Levan, Albert; Fredga, Karl, and Sandberg, Avera A.: Nomenclature for centromeric position on chromosomes. *Hereditas, 52:* 201, 1964.

Lieppman, Michael, and Hubbs, Clark: A karyological analysis of two cyprinid fishes, *Notemigonus crysoleucas* and *Notropis lutrensis.* Texas reports on Biology and Medicine, 27 (No. 2): 427, 1969.

Makino, Sajiro: *Chromosome Numbers in Animals.* Ames, Iowa, Iowa State College Press, 1951.

Maio, J.J., and Schildkraut, C.L.: Isolated mammalian metaphase chromosomes. I. General characteristics of nucleic acids and proteins. *J. Molec. Biol., 24:* 29, 1967.

Mayers, Lyndon J., and Roberts, Franklin L.: Chromosomal homogeneity of five populations of alewives, *Alosa pseudoharengus. Copeia,* No. 2: 313, 1969.

McPhail, J.D., and Jones, R.L.: A simple technique for obtaining chromosomes from teleost fishes. *J. Fish, Res. Bd. Canada, 23:* 5, 1966.

Muramoto, Jun-Ichi, and Ohno, Susumu: On the diploid state of the fish order Ostariophysi. *Chromosoma, 24:* 59, 1968.

Nogusa, Shynsaku: Chromosome studies in pisces V. Variation of the chromosome number in Acheilognathus rhombea due to multiple-chromosome formation. *Annotationes Zoologicae Japonenses, 28* (4): 249, 1955.

Nogusa, Shyunsaku: A comparative study of the chromosomes in fishes with particular considerations on taxonomy and evolution. *Memoirs of the Hyogo University of Agriculture,* 3 (1) : 1, 1960.

Ohno, S.; Stenius, Faisst E., and Zenzes, M.T.: Post-Zygotic chromosomal rearrangements in rainbow trout *(Salmo irideus* Gibbons). *Cytogenetics,* 4: 117, 1965.

Ohno, Susumu, and Atkin, N.B.: Comparative DNA values and chromosome complements of eight species of fishes. *Chromosoma,* 18: 455, 1966.

Ohno, Susumu; Buramoto, Junichi, and Christian, Lawrence: Diploidtetraploid relationship among old-world members of the fish family Cyprinidae. *Chromosoma,* 23: 1, 1967.

Ohno, Susumu; Wolf, Ulrich, and Atkin, Niels, B.: Evolution from fish to mammals by gene duplication. *Hereditas,* 59: 169, 1968.

Ohno, S.; Muramoto, J., and Christian Stenius L.: Microchromosomes in Holocephalian, Chondrostean and Holostean fishes. *Chromosoma,* 26: 35, 1969.

Ojima, Y., and Hitotsumachi, S.: Cytogenetic studies in lower vertebrates. IV. A note on the chromosomes of the carp *(Cyprinus carpio)* in comparison with those of the funa and the goldfish *(Carassius auratus). Japan. J. Genetics, 42* (3): 163, 1967.

Pegington, C.J., and Rees, H.: Chromosome size in salmon and trout. *Chromosoma,* 21: 475, 1967.

Prokofieva, A.: On the chromosome morphology of certain pisces. *Cytologia,* 5: 498, 1934.

Priest, Jean H.: *Cytogenetics.* Philadelphia, Lea and Febiger, 1969.

Rachlin, J. W.; Perlmutter, A., and Seeley, R.J.: Monolayer culture of gonadal tissue of the zebra danio *Brachydanio rerio. Progressive Fish Culturist, 29:* 232, 1967.

Rees, H.: The question of polyploidy in the salmonidae. *Chromosoma, 15:* 275, 1964.

Rees, H.: The chromosomes of *Salmo salar. Chromosoma, 21:* 472, 1967.

Regan, James D.; Sigel, Michael M.; Lee, William H.; Llamas, Kirsten A., and Beasley, Annie R.: Chromosomal alterations in marine fish cells in vitro. *Can. J. Genet. Cytol., 10:*448, 1968.

Rendel, J.M.: The effects of genetic change at different levels, In: *Ideas in Modern Biology,* J.A. Moore (Ed.). N.Y., Natural History Press, 1965.

Roberts, Franklin L.: A chromosome study of twenty species of Centrarchidae. *J. Morpho., 115:* 401, 1964.

Roberts, Franklin L.: Chromosome cytology of the Osteichthyes. *The Progressive Fish-Culturist,* April: 75, 1967.

Roberts, Franklin L.: Chromosomal polymorphism in North American land-locked *Salmo salar. Can. J. Geent. Cytol., 10:* 865, 1968.

Roberts, Franklin L.: Atlantic salmon *(Salmo salar)* chromosomes and speciation. *Trans. Amer. Fish. Soc.,* No. 1: 105, 1970.

Schultz, Jack R.: Gynogenesis and triploidy in the viviparous fish *Poeciliopsis. Science, 157:* 1564, 1967.

Simon, Raymond C.: Chromosome morphology and species evolution in the five North American species of Pacific salmon *(Oncorhynchus). J. Morphol., 112* (1): 77, 1963.

Simon, R.C.: Fixation and fat extraction before staining and squashing, for chromosomes of fish embryos. *Stain Technol.,* 39: 45, 1964.

Sterba, Gunther: *Freshwater Fishes of the World,* N.Y., Pet Library LTD., 1962.

Stewart, Kenneth W. and Levin, Catherine B.: A method of obtaining permanent dry mounted chromosome preparations from teleost fish. *J. Fish. Rec. Bd. Canada, 25* (5): 1091, 1968.

Subrahmanyam, K.: A Karyotypic study of the estuarine fish *Boleophthalmus boddaeri* (Pallas) with calcium treatment. *Curr. Sci., 38* (18): 437, 1969.

Svardson, G.: Chromosome studies on Salmonidae. *Medd. St. undersokn. forsoksanst. Sotvattensfisket., 23:* 1, 1945.

Swanson, Carl P.: *Cytology and Cytogenetics.* N.J., Prentice-Hall, 1957.

Taylor, J.H.: *Am. Naturalist, 91:* 209, 1957.

Taylor, Kenneth M.: The chromosomes of some lower chordates. *Chromosoma, 21:* 181, 1967.

Uyeno, Teruya, and Miller, Robert Rush: Multiple sex chromosomes in a Mexican cyprinodontidontid fish. *Nature, 231;* 452, 1971.

Uyeno, T., and Miller, R.R.: Second discovery of multiple sex chromosomes among fishes. *Experientia, 15:* 223, 1972.

Uyeno, Teruya, and Smith, G.R.: Tetraploid origin of the karyotype of catostomid fishes. *Science, 175:* 644, 1972.

Wolf, Ken, and Quimby, M.C.: Fish cell and tissue culture. In: Hoar, W.S. and Randall, D.J. (Ed.) : *Fish Physiology,* Vol. 3, N.Y., Academic Press, 1969.

Yamamoto, T.: Progenies of induced sex-reversal females mated with induced sex-reversal males in the medaka, *Oryzias latipes. J. Exptl. Zool.,* 146: 163.

Yamamoto, Toki-O: Sex differentiation In: *Fish Physiology,* W.S. Hoar (Ed.). N.Y., Academic Press, 1969.

APPENDIX

STAINS

Aceto-orcein (2 percent)

1. Dissolve 2 grams of orcein stain (natural or synthetic) in 45 ml of hot glacial acetic acid and boil gently (under a hood) for five minutes with stirring.
2. Filter while hot, allow to cool, and store in a refrigerator. Before use, warm to room temperature, shake well, and filter the amount needed for staining.

Giemsa Stain (May-Grunwald)

Stock Solution

Giemsa powder .... 1.0 gram
Glycerin ....66.0 ml
Methanol ....66.0 ml

Mix glycerin and Giemsa powder, heat in oven at 60°C for two hours, add methanol.

Working Solution

Dilute 1 ml of stock Giemsa with 20 ml of distilled water. Always make fresh. Do not re-use.

Feulgen Staining

Leuco-Basic Fuchsin (Schiff's reagent)

1. Dissolve 1 gram of basic fuchsin in 200 ml of boiling distilled water and cool to 60°C.
2. Filter and add 2 grams of sodium metabisulfite and 10 ml HCl to the filtrate.
3. Shake well and store in tightly stoppered, dark bottle for twenty-four hours.
4. Add 0.5 grams of fresh activated charcoal (e.g. Norite) and shake for one to two minutes.
5. Filter rapidly through *dry* coarse filter paper. Store in a dark bottle in refrigerator.

Sodium Metabisulfite

Mix 5 ml of 10 percent sodium metabisulfite with 5 ml of 1 N HCl and 100 ml of distilled water.

Procedure

1. Rinse slides containing air-dried preparations in distilled water and hydrolyze in 1 N HCl at 60°C for 4 to 8 minutes.
2. Stain in leuco-basic fuchsin for 1 to 2 hours.
3. Rinse in distilled water and add to sodium bisulfite solution for 3 changes of 5-10 min each. Rinse, air dry and mount. The DNA is seen in shades of reddish-purple.

Chromosome Fixative

Thoroughly mix 15 ml of absolute methyl alcohol with 5 ml of glacial acetic acid immediately before each use. Always make fresh and never re-use.

PURCHASING MATERIALS (U.S. ONLY)

Grand Island Biological Co. (3175 Staley Road, Grand Island, N.Y. 14072 and 2323 fifth street, Berkeley, California 94710).

Phytohemagglutinin (M form, lyophilized)
Pokeweed mitogen
Colcemid
Antibiotics
Tissue culture media
Aceto-orcein staining solutions (1% and 2%)
Trypan blue
Velban
Trypsin
Serums
Fish cell lines (Fathead minnow and Rainbow trout gonad)
Extracts

Fisher Scientific Company (Central office at 711 Forbes Avenue, Pittsburgh, Pennsylvania 15219).

Orcein stain (synthetic)
Giemsa stain
Fuschin
Sodium metabi ulfite
Kronig's cement

Sigma Chemical Company (P.O. Box 14508, St. Louis, Missouri. 63178

Colchicine

PURCHASING SPECIMENS

The following list is but a few of the many possible sources where fishes may be ordered.*

Gulf Specimen Company, Inc. (P.O. Box 237 Panacea, Florida 32346)
Charles P. Chase Co., Inc. (7330 N.W. 66th St. Miami, Florida. 33166)
Culf Coast Worm Gardens (3125 15th Ave. N. Texas City-La Marque, Texas 77590).
The Harborton Marine Laboratory (Box 11 Harborton, Virginia 23389).
The Lemberger Co. (1222 W. South Park Ave. Oshkosh, Wisconsin 54901).
Paramount Aquarium-Ertrachter Fish Farm (Rte. 5, Box 5 Tampa, Florida 33614).
Pacific Bio-Marine Supply Co. (P.O. Box 536 Venice, California 90291).
E. G. Steinhilber and Co., Inc. (102 Josslyn St. Oshkosh, Wisconsin 54901).
Tarpon Zoo, Inc. (P.O. Box 847 Tarpon Springs, Florida 33589).
Truslow Farms Incorporated (Chestertown, Maryland 21620).
Zoological Center, International (15W506 W. 63rd (Burr Ridge) Hinsdale, Illinois 60521).

*Names and Addresses taken from *Animals for Research*. Publication 1678. National Academy of Sciences. Washington, D.C., 1968.

GENERIC INDEX

SUBJECT INDEX